THE CARIBBEAN ISLANDS

TOURING NORTH AMERICA

SERIES EDITOR
Anthony R. de Souza, *National Geographic Society*

MANAGING EDITOR
Winfield Swanson, *National Geographic Society*

THE CARIBBEAN ISLANDS

Endless Geographical Diversity

BY

THOMAS D. BOSWELL

AND

DENNIS CONWAY

RUTGERS UNIVERSITY PRESS • NEW BRUNSWICK, NEW JERSEY

#25547065

This book is published in cooperation with the 27th International Geographical Congress, which is the sole sponsor of *Touring North America*. The book has been brought to publication with the generous assistance of a grant from the National Science Foundation/Education and Human Resources, Washington, D.C.

Rutgers University Press
109 Church Street
New Brunswick, New Jersey 08901

The paper used in this book meets the minimum requirements of American National Standard for Information Sciences—Permanence of Paper for Printed Library Materials, ANSI Z39.48-1984.

Library of Congress Cataloging-in-Publication Data

Boswell, Thomas D.
    The Caribbean Islands: endless geographical diversity / Thomas D. Boswell and Dennis Conway.—1st ed.
        p.   cm.—(Touring North America)
    Including bibliographical references and index.
    ISBN 0-8135-1894-6 (cloth)—ISBN 0-8135-1895-4 (pbk.)
    1. West Indies—Description and travel—1981–   —Guidebooks. I. Conway, Dennis, 1941–   . II. Title. III. Series.
F1609.B6   1992
917.2904M′52—dc20   92-11579
                                                                                CIP

First Edition

Frontispiece: El Yunque National Forest, Puerto Rico. Photograph by Thomas D. Boswell.

Series design by John Romer

Typeset by Peter Strupp/Princeton Editorial Associates

△ Contents

△ Foreword

Touring North America is a series of field guides by leading professional authorities under the auspices of the 1992 International Geographical Congress. These meetings of the International Geographical Union (IGU) have convened every four years for over a century. Field guides of the IGU have become established as significant scholarly contributions to the literature of field analysis. Their significance is that they relate field facts to conceptual frameworks.

Unlike the last Congress in the United States in 1952, which had only four field seminars all in the United States, the 1992 IGC entails 13 field guides ranging from the low latitudes of the Caribbean to the polar regions of Canada, and from the prehistoric relics of pre-Columbian Mexico to the contemporary megalopolitan eastern United States. This series also continues the tradition of a transcontinental traverse from the nation's capital to the California coast.

This guide takes us through the Caribbean to the richly diverse physical and cultural landscapes of the islands of Puerto Rico, Tortola, St. Martin, Guadaloupe, St. Lucia, Antigua, St. Thomas, Grenada, Trinidad, Barbados, and Martinique. Island visits disclose colonial imports of the Spanish, Dutch, French, British, and Americans as revealed through the periods of slavery and the plantation economies to the modern period of tourism and national identification.

Professors Thomas D. Boswell (University of Miami) and Dennis Conway (Indiana University) are the leading scholars on the geography of the Caribbean.

Anthony R. de Souza
BETHESDA, MARYLAND

△ Acknowledgments

The production of this guidebook involved the efforts of many people, to all of whom we are deeply indebted. Thomas Boswell particularly wants to express his gratitude to Dr. Angel David Cruz Báez (Department of Geography, University of Puerto Rico), Dr. James Keech (Interamerican University and Ft. Buchanan High School, San Juan, Puerto Rico), Dr. Frank L. Mills (Eastern Caribbean Center, University of the Virgin Islands), Mrs. Doris Jadan (Cruz Bay, St. John, U.S. Virgin Islands), Dr. Francois Van Der Hoeven (Netherlands Antilles National Parks Foundation, Sint Maartin), Mr. Ivor B. Ford (Ministry of Education, Antigua), Mr. Alfred Christopher (former Minister of Education, Tortola, British Virgin Islands), Mr. Sydney Braithwaite (formerly of the Ministry of Education, Tortola, British Virgin Islands), Dr. John P. Augelli (Department of Geography, University of Kansas), and Dr. Frank Innes (Department of Geography, University of Windsor, Canada).

Dennis Conway expresses his thanks to Dr. Jean-Pierre Chardon (Port Autonome, Point-a-Pitre, Guadeloupe), Mr. Robert Devaux, Mr. Leonard Waite, and Mr. Giles Romulus (The St. Lucia National Trust, Casteries, St. Lucia), Mr. Eugene Pilgrim and Mrs. Nola Myers (Barbados Geographical Association, Barbados), Mr. Desmond John (Population Planning Association, St. George's, Grenada), and Mrs. Diana Clarke (Classic Tours, Port of Spain, Trinidad and Tobago).

We acknowledge the dedicated work of the following cartographic interns at the National Geographic Society, who were responsible for producing the maps that appear in this book: Nikolas H. Huffman, cartographic designer for the 27th IGC; Patrick Gaul, GIS specialist at COMSIS in Sacramento, California; Scott Oglesby, who oversaw production of the shaded relief artwork; Lynda Barker; Michael B. Shirreffs; and Alisa Pengue Solomon. The shaded relief in this volume was drawn by Michael Shirreffs. Assistance was provided by the staff at the National Geographic Society, especially the Map Library and Book Collection, the Illustrations Library, the Cartographic Division, Computer Applications, and Typographic Services. Special thanks go to Susie Friedman of Computer Applications for procuring the hardware needed to complete this project on schedule.

We also thank Lynda Sterling, publicity manager and executive assistant to Anthony R. de Souza, the series editor; Richard Walker, editorial assistant at the 27th International Geographical Congress; geography interns at the National Geographic Society Natalie Jacobus, who proofread the volume, and Tod Sukontarak, who indexed the volume and served as photo researcher. They were major players behind the scenes. Many thanks, also, to all those at Rutgers University Press who had a hand in the making of this book—Kenneth Arnold, Karen Reeds, Marilyn Campbell, and Barbara Kopel.

Errors of fact, omission, or interpretation are entirely our responsibility, and any opinions and interpretations are not necessarily those of the 27th International Geographical Congress, which is the sponsor of this field guide and the *Touring North America* series.

This guidebook is dedicated to Dr. Robert A. Lewis, Professor Emeritus of Geography, Columbia University; Dr. Abram J. Jaffe, Professor Emeritus of Sociology, Columbia University; Dr. Kingsley E. Haynes, Professor and Dean of Graduate Studies, George Mason University; and Dr. Ronald Briggs, Professor and Director of Computer Services, University of Texas at Dallas.

The authors' division of labor was as follows: Thomas D. Boswell wrote the overview of the Caribbean in Part One and the Hints to the Traveler. He also wrote the descriptions of Puerto Rico, the Virgin Islands, Sint-Maarten and Saint-Martin, and Antigua in Part Two. Dennis Conway wrote the descriptions of Barbados, the French Antilles, St. Lucia, Grenada, and Trinidad and Tobago in Part Two. Both authors contributed to the Suggested Readings.

PART ONE

Introduction to the Region

Regions of the Caribbean

△ Introduction

Probably no place in the world of similar size exhibits as much diversity as the islands of the Caribbean. This diverseness manifests itself in virtually every aspect of the islands making up this region, including its landforms, climate, vegetation, human history, political systems and degree of independence, levels of economic development, religious preferences, and even the languages spoken by its citizens. In short, it is a wonderful laboratory for illustrating basic geographical principles, in addition to being an idyllic location for a tropical vacation.

The thousands of islands, islets, keys, and rocks that comprise the archipelago of the Caribbean stretch approximately 2,200 miles (3,500 kilometers) from Cape San Antonio on the western tip of Cuba eastward through the islands of Hispaniola, Puerto Rico, and the Virgin Islands and then southward through the Lesser Antilles to the northern coast of South America. The aggregate land area of the islands included in this region is almost 92,000 square miles (238,000 square kilometers), slightly less than the area of the United Kingdom or the state of Oregon in the United States.

CARIBBEAN TOPONYMS

The terminology used to describe these many and diverse islands is truly bewildering, being a reflection of the diverse cultures that have inhabited them. This is important to note because there is no unanimity or consensus regarding how some of the toponyms of this region should be used, either by the residents of these islands or by the scholars who study them. For instance, we use the regional term Caribbean in this book to include not only the islands in the Caribbean Sea, but also the islands comprising the Commonwealth of the Bahamas and the British possession of the Turks and Caicos Islands, east of Florida and north of Cuba in the Atlantic Ocean. Ours is an islands' connotation, but other writers additionally include the countries that encompass the Caribbean Sea, such as most of the countries of Central America and the republics of northern South America.

It is logical to divide the Caribbean islands into five subregions. The largest is the Greater Antilles, encompassing the four islands of Cuba, Hispaniola, Jamaica, and Puerto Rico and the five political entities Cuba, the Dominican Republic, Haiti, Jamaica, and Puerto Rico. Together these islands include eighty-nine percent of the land area in the Caribbean (see Appendix). The name "Antilles" derives from "Antilia," the mythical land in the Atlantic Ocean between Europe and the Orient. The second subregion is the Lesser Antilles, including the smaller islands extending from the Virgin Islands east of Puerto Rico and southward to Trinidad. It comprises an additional four percent of the area of the Caribbean islands. The Bahamas and Turks and Caicos Islands comprise the third subregion and account for almost six percent of the land area of the Caribbean. The remaining two subregions collectively comprise less than one percent of the area of the Caribbean. These are the Cayman Islands, south of Cuba and east of Jamaica, and the Dutch ABCs (of Aruba, Bonaire, and Curaçao). Hundreds of other islands are technically located in the Caribbean, but are possessions of Central American and northern South American countries, and most geographers consider them part of Latin America instead of the Caribbean.

Several other terms that require definition are used widely throughout the Caribbean. Perhaps the most important of these is the phrase West Indies. When Columbus reached the Caribbean during his first voyage in 1492 he thought he had discovered a shorter western route to the Orient and he called these islands "Las Indias," and the natives who lived on them became known as "Indians." Later, when it was realized that these were not the same islands as those located in Southeast Asia, they became know as the West Indies, to distinguish them from the East Indies. During the 400-year period between the fifteenth and nineteenth centuries, the West Indies were settled mainly by the Spanish, French, Dutch, and British colonialists. This complicated the terminology because the Spanish territories of Cuba, the Dominican Republic, and Puerto Rico became part of Latin America. Today, most local residents consider only the non-Hispanic islands of the Caribbean to comprise the West Indies.

The West Indies also are subdivided according to their colonial history. Thus, the Netherlands Antilles include the Dutch possessions of Aruba, Bonaire, Curaçao, St. Eustatius, Saba, and the southern half of Sint Maarten. The French West Indies include the French-owned islands of Martinique, Guadeloupe, the northern half of St. Martin, St. Barthélemy, and sometimes Haiti (which became independent in 1804). The British West Indies comprise those many islands that had a British colonial history. In addition, the U.S. Caribbean includes Puerto Rico and the U.S. Virgin Islands.

The Lesser Antilles are frequently subdivided by English- and French-speaking residents into the Leeward Islands and the Windward Islands. The Dominica Passage between the islands of Guadeloupe and Dominica is usually regarded as the dividing line, with those to the north called the Leewards and those to the south the Windwards. Interestingly, there is no climatological logic for this designation because these islands are similar in terms of their exposures to the prevailing northeastly Trade Winds. One explanation is that, during his second voyage in 1493, Columbus sailed westward between the islands of Guadeloupe and Dominica to be on the leeward side of the Lesser Antilles, thereby gaining some protection from hurricanes that blew into the area from the east. To complicate matters, the Dutch call their northern islands of St. Maarten, Saba, and St. Eustatius the Windwards, and the ABCs to the south the Leewards, just the reverse of British and French use of these terms. The reason for the Dutch terminology is that Aruba, Bonaire, and Curaçao—off the coast of Venezuela—are much drier than their three northern cousins.

LANDFORMS

The diversity of the Caribbean is reflected in its many types of landforms. Some islands are mountainous, while others are virtually flat. A few are volcanically active and many experience frequent and devastating earthquakes, but several (like most in the Bahamas) have not felt any kind of earth movement for hundreds of years. Elevations range from a high of 10,417 feet (3,175 meters) at Pico Duarte in the Dominican Republic to about –150 feet (–46 meters) in the southern Enrequillo Depression in the same country. But the true range in elevations is much greater than suggested by these figures if the depth of the Puerto Rican Trench, about 40 miles (64 kilometers) off the north coast of Puerto Rico, is considered, because it plunges more than 30,000 feet (9,146 meters) below sea level, deeper than Mount Everest is high. For the most part, the Caribbean is an extraordinarily active geologic region. The volcanic activity and earthquakes that typify this area reflect its location at the site of five large, colliding tectonic plates. At the center of it all is the Caribbean Plate, whose northern and eastern edges contain most of the islands in the Caribbean. The most violent collision is occurring along the eastern edge of the Caribbean Plate where the westward-moving Atlantic Plate is plunging under it at a rate of between one and two inches (three to four centimeters) per year. As the more dense Atlantic Plate dives under the lighter Caribbean Plate, the leading edge of the latter deforms, creating a crumpled mountainous landscape with consequent volcanic activity.

The geologic details of the Lesser Antilles indicate that they are comprised of two roughly parallel arcs. The innermost is volcanic in origin and more mountainous. It is the one that includes most of the larger islands in this archipelago, such as St. Kitts, Nevis, Montserrat, Dominica, Martinique, St. Lucia, St. Vincent, and Grenada. The outer arc of islands is more low-lying, being of limestone origin or deriving from much older volcanic activity that is no longer taking place. Examples of the latter include the Virgin Islands, Anguilla, St. Martin, Barbuda, Antigua, and Barbados.

The two halves of the island of Guadeloupe contain examples of both inner and outer island arc characteristics. The southwestern half, known as Basse-Terre, is a volcanic island whose highest peak (Mount Soufrière) reaches an elevation of 4,812 feet (1,467 meters). Conversely, its northeastern half, Grande-Terre, is a low-lying, undulating island of limestone composition. In addition to Guadeloupe's Basse-Terre, evidence of current volcanic activity is also found on Martinique and St. Vincent. Several other islands such as Montserrat, Nevis, St. Lucia, Redonda, and Dominica contain hot sulfur springs, reflections of dying but recent vulcanism.

The Greater Antilles are products of much older and probably stronger tectonic forces dating mainly to between 60 million and 90 million years ago. As a consequence they are not typified by the recency of volcanic activity that characterizes the Lesser Antilles. But when they were exposed to uplifting and vulcanism they experienced mountain-building that was even greater than that in the eastern Caribbean. As a consequence, the highest mountains on Cuba, Jamaica, and Hispaniola are taller than those on the Lesser Antilles. Even on Puerto Rico, the smallest of the Greater Antilles, the highest elevation of 4,389 feet (1,338 meters) at Cerro de Punta is about equal to the highest peak in the eastern Caribbean, Guadeloupe's Mount Soufrière (4,812 feet or 1,467 meters). The main difference in the tectonics of the Greater and Lesser Antilles is that vulcanism has ceased on the four largest islands, even though they still occasionally experience devastating earthquakes. For instance, Port Royal, the old capital of Jamaica, was completely destroyed by an earthquake in 1692. The current capital, Kingston, was similarly destroyed in 1907 and badly damaged again as recently as 1957. Port-au-Prince, capital of Haiti, was destroyed twice during the nineteenth century, as was its northern sister city, Cap Haitien, in 1842.

The reason for the occurrence of earthquakes, but not volcanic activity, in the Greater Antilles is also explained by the principles of plate tectonics. Whereas in the eastern islands the Caribbean and Atlantic plates collide head-on, the movement of the Caribbean and North American plates is horizontal to each other. They slide past each other, rather than one diving under the other. Thus,

Caribbean Region Tectonics

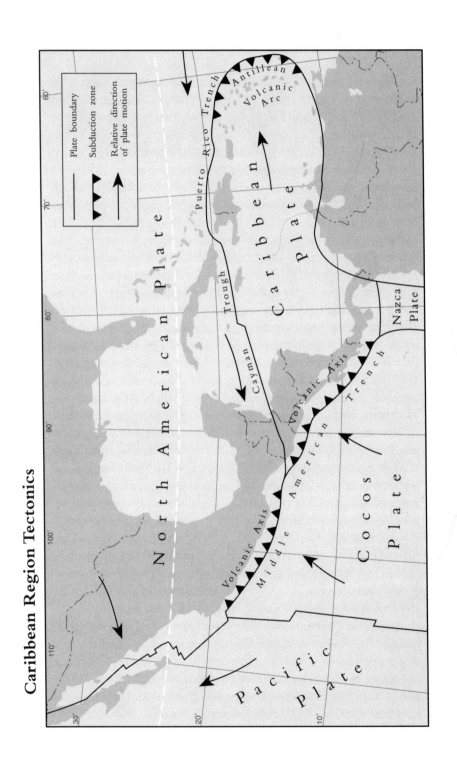

the force of contact is less intense along the northern coasts of the four larger islands. The friction generated causes pressure to build, and when it is released this causes the infrequent earthquakes that affect all of the northern islands from Cuba in the west to Antigua in the Lesser Antilles to the east. In fact, both the Puerto Rican and Cayman trenches were created by this heavy frictional movement, as well as by earlier subduction forces when the Caribbean and North American plates confronted each other more directly than they do today.

The ancient vulcanism that created the Greater Antilles and some of the older outer arc of the Lesser Antilles was followed by a period of submersion beneath the sea. During this period of several million years, thick layers of limestone were deposited by coral formations and dying calcarious marine organisms. As a result, many of these islands have limestone soils and rocks that cover the lower elevations of their outer edges. The limestone deposits on the higher elevations of these islands were quickly stripped away when another uplift occurred, raising them once again above sea level. The greater steepness of the slopes at higher elevations intensified the erosive effects of the tropical rainfall on the limestone covering, but the inner volcanic cores of the uplands were more resistant to this process of degradation. The alluvium washed from the mountains was deposited in the lower valleys and the narrow coastal plains where the limestone accumulations are most evident today. On the other hand, the older mountainous cores are composed of igneous rocks, evidence of their volcanic origins. However, the mountains on these islands no longer exhibit the symmetrical shape typical of youthful cinder cones because they have been eroded for so many millions of years.

Several of the low-lying islands, such as those in the Bahamas, the Turks and Caicos Islands, Anegada in the British Virgin Islands, and the Cayman Islands are products primarily of coral reefs that grew on top of submerged sea platforms called banks. They were either uplifted above sea level or covered by blowing sand and dying marine organisms which accumulated on top of them. These islands tend to appear to be nearly level; maximum elevations rarely exceed 200 feet (61 meters).

In some places the thick coastal layers of limestone have weathered into geologic formations known as karst. Here water has dissolved the underlying calcarious materials into caves with spectacular underground features such as stalactites and stalagmites (spines hanging from the ceiling and columns built up from the ground, respectively). In some places rivers disappear underground, sinkholes abound, and the surface takes on a jumbled appearance of depressions and low rounded hills known as haystacks. These features are particularly prevalent in parts of all four of the Greater Antilles, especially on their wetter northern coasts, and on Grande-Terre in Guadeloupe. Limestone caves are also present in less spectacular fashion on many of the other smaller islands, as are a variety of sinkholes, known in the British Caribbean as blue holes because deep sea-water intrusion makes them look blue.

CLIMATE AND WEATHER PATTERNS

Taxi drivers in the Caribbean are amused by one of the questions tourists from North America and Europe invariably ask when they arrive in the Caribbean: "How has the weather been?" The truth is the day-to-day weather is almost always the same at any given location, a characteristic of most tropical locations in the world. This, however, is not to suggest that there are not variations between places within the Caribbean. There are marked differences, particularly in the amount of annual rainfall. For instance, drier areas are the relatively flat islands (like the Dutch ABCs), the valleys shielded by mountains from moderating sea breezes (like the Enrequillo Depression in the Dominican Republic), and the leeward sides of the Greater Antilles (like the southeastern coast of Jamaica). These drier areas typically have a somewhat greater diurnal range in temperature, tending to daily highs that are a few degrees higher and lows that are three or four degrees lower.

Five factors primarily influence the climate conditions of the Caribbean: latitude, prevailing winds, the presence of a large mass of water, the location of mountains, and the occurrence of tropical cyclonic storms.

Latitude

With the exception of the northern two thirds of the Bahamas, all of the Caribbean is in the tropics north of the Equator. It is this factor that largely controls the uniform temperatures that typify this area. In the tropical latitudes the angle of the sun's rays and the length of the daylight period do not vary here the way they do in the middle and higher latitudes. The variations in average temperatures for the hottest and coolest months here are almost always less than 10°F (6°C), a range that decreases closer to the Equator. For instance, the annual range of temperatures between the coldest and warmest months for San Juan, Puerto Rico, is 6°F (3.3°C); for Fort-de-France, Martinique, it is 5°F (2.8°C); for Bridgetown, Barbados, it is 4°F (2.2°C); and for Port of Spain, Trinidad, it is 4°F (2.2°C). Even for Nassau, in the Bahamas, just north of the Tropic of Cancer, the temperature range is only about 12°F (6.7°C). Because the differences between the winter and summer temperatures is slight, residents of the Caribbean like to joke that nighttime is the true winter of the islands. In fact, the major difference between winter and summer seasons here is in the amount of rainfall, not temperature. Summer is the wet season and winter is the drier period. Daily highs year-round usually range between 82°F (28°C) and 92°F (33°C), whereas the lows range between 70°F (21°C) and 80°F (27°C). Exceptions occur on the rare occasions when a cold north wind (called a Norte) blows south into the Caribbean from the interior of North America, but these incursions only affect temperatures of the Greater Antilles and the Bahamas.

Prevailing Winds

The islands of the Caribbean lie in the belt of the Northeast Trade Winds, so-called because they blow from the northeast toward the southwest and because of the role they played in early European settlement and trading patterns in this region. From a climatological perspective they are significant because their direction determines the leeward and windward sides of the islands, which affects the amount of precipitation and modifies somewhat the temperature conditions on opposite sides of the islands. Generally, the windward coasts are wetter and have somewhat less extreme temperatures. But the latter is only a minor effect. Usually, daily highs on the windward coasts range between 2°F and 4°F (1 to 2°C) lower than on the leeward side of the islands.

Water Mass

Being surrounded by water has two important climatological consequences for the Caribbean Islands. First, the large body of water provides a virtually limitless source of evaporation for the warm tropical Northeast Trade Winds that blow over this area, increasing the humidity of Caribbean air masses. Second, because water both heats and cools more slowly than land it helps moderate the temperatures of the islands, especially in coastal areas. Except for a few sheltered locations, like the Cul de Sac depression of Haiti and the Valle Central of Cuba, temperatures as high as a 100°F (38°C) and as low as 60°F (16°C) are rare in the Caribbean.

Mountains

Air masses in the Caribbean are normally both warm and humid. Whether rainfall occurs depends upon providing a mechanism for forcing this air to rise, thereby cooling it enough to create condensation that will precipitate. One such mechanism is the presence of mountains on the islands. Air is forced to rise on their windward sides, while it subsides on the leeward sides. As a consequence, the windward sides of the mountains normally experience heavy rainfall (locally called relief rainfall), while the leeward sides are relatively dry, lying in the rain-shadow of the mountains. These windward and leeward effects are most fully developed in the Greater Antilles, where the mountains are higher and distances traveled by air masses greater. For example, the city of Port Antonio on the northeastern coast of Jamaica receives an average of 125 inches (3,172 millimeters) of annual precipitation. The Blue Mountains 20 miles (32 kilometers) inland get about 200 inches (5,076 millimeters), and Kingston (35 miles or 56 kilometers southwest of Port Antonio) on the southeastern coast experiences only 30 inches (761 millimeters) of rain per year. In the smaller mountainous islands in the Lesser Antilles, rainfall tends to be more symmetrical without major differences between the windward eastern coasts and the leeward western sides. For instance, the coastal lowlands of both the eastern and western sides of St. Vincent average about 70 inches (1,875 millimeters) of rainfall, while the interior mountains receive more than 125 inches (3,173 millimeters). The flatter islands, like the Bahamas and Caymans, receive much less rainfall than the mountainous ones. Most of the Bahamian islands receive between 40 and 50 inches (100 to 125 centimeters) of rain, while the Dutch ABCs receive only between 20 and 30 inches (50 to 75 centimeters) per year. In fact, availability of water for drinking, agriculture, industry, and urban uses is of major concern in the Caribbean. Much of the rain that does fall runs off to the surrounding seas before it can be used. Some of the islands now strictly ration water. Houses are frequently built with rain-catching roofs and cisterns and many of the drier islands have desalination plants. Signs in hotel rooms urge patrons not to waste water.

A typical summer day on an island in the Caribbean will begin with a cool clear morning with temperatures around 75°F to 80°F (24 to 27°C). During the day temperatures increase. By the late afternoon they warm to between 87°F (31°C) and 92°F (33°C) and a gentle onshore breeze develops, as convectional heating creates a low pressure system over the land, relative to somewhat higher pressure over the adjacent and cooler sea. Late afternoon convectional showers are intense but of short duration (lasting less than an hour). The early evenings are warm and muggy because of the high relative humidity. Later at night (after about 10 p.m.) the earlier daytime onshore breeze reverses itself to a gentle offshore wind because a weak, high pressure cell develops over the land as its temperatures drop. Frequently, spectacular lightning flashes can be seen over the warmer adjacent sea as rain falls there. During the winter, temperatures are only slightly lower, but significantly less rain falls. Although there are variations from place to place, normally about sixty-five to seventy-five percent of the precipitation occurs during the summer, between May and October.

Tropical Storms

Tropical cyclonic storms develop most often between the tenth and twenty-fifth parallels in the northern hemisphere. Although one did occur there in 1933, Trinidad is the only island that is almost immune to them because it is the southernmost of the Caribbean islands. Low pressure systems often develop east of the Lesser Antilles in the middle Atlantic Ocean. Current wisdom suggests that some of the more fully developed lows originate as thermal bubbles over the western part of Africa. As these thermals travel westward toward the Caribbean they may mature into large and destructive storms, if conditions are right. They start out as tropical disturbances. When their sustained winds reach 25 miles (40 kilometers) per hour they become known as depressions. If they reach a strength of 39 miles (63 kilometers) per hour they are called storms and given a name. If they increase in size and strength to have winds of 74 miles (119 kilometers) per hour or more they become hurricanes.

During a typical hurricane season in the Caribbean, which lasts officially from June 1 to November 30, perhaps fifty to sixty tropical depressions will develop, but only between ten and fifteen will evolve into tropical storms, and fewer still (between three and seven) will mature into hurricanes. The depressions and storms are both notable because of the rain they generate, but it is the hurricanes that are most destructive. A hurricane's sustained winds can reach more than 150 miles (242 kilometers) per hour, with gusts exceeding 200 miles (322 kilometers) per hour. Even the minimal winds of 74 miles (119 kilometers) per hour can easily knock down a person. Such heavy wind destroys crops, trees, and houses, and kills people. But wind is not the only destructive element of a hurricane. The heavy rainfall causes massive flooding and pollutes drinking water supplies by overflowing sewage systems. In addition, virtually all hurricanes are accompanied by a tidal surge from 5 to 20 feet (1.5 to 6 meters) in height, which always causes massive destruction along coastal areas. A little-known fact (outside the Caribbean) is that small tornados also accompany virtually all hurricanes and add their devastating effects. Normally, these violent storms follow an east-to-west track through the Caribbean and then curve northward toward the United States, but the mechanisms that steer these storms are not well understood.

Although use of satellites and reconnaissance aircraft have improved early detection of hurricanes and their warning to local residents, their paths still are unpredictable and each year millions of dollars of property damage and often numerous deaths occur. In 1988 Hurricane Gilbert caused massive destruction in the Caribbean. With sustained winds of 175 miles (280 kilometers) per hour, it entered the Caribbean by passing over St. Lucia. It then headed for the Dominican Republic and Haiti. Afterward it continued westward to Jamaica and traveled the entire length of that island, rendering 500,000 people homeless and ruining 60 percent of the island's coffee crop and destroying most of its bananas. Finally it headed for Mexico's Yucatán Peninsula, where it did major damage to the resort areas of Cancún and Cozumel before it finally died. In its path, damage exceeded $200 million, hundreds of deaths, and hundreds of thousands of homes destroyed or badly damaged.

VEGETATION PATTERNS

Vegetation patterns in the Caribbean vary according to rainfall, temperature, soil, and human influence. Along rocky and sandy coastlands it is common to find low bushes and sea grapes, often bent low and sometimes flattened by the strong prevailing Trade Winds. In brackish, swampy, and coastal areas, clumps of mangroves often grow. The mangrove trees have a durable and tolerant root system that provides a measure of protection against the erosive effects of the sea. In fact, their dense and thick root system helps accumulate mud and sand from shallow coastal areas, extending the land seaward. Also, common in sandy coastal areas are coconut palms.

Tropical rain forests were originally found on the wetter northern and eastern coasts of the islands, to an elevation of about 3,000 feet (915 meters). Unfortunately, most of the original cover of this vegetation has been removed by people clearing land for agriculture, especially in the lowlands and on lower slopes. Still, rain forests are found on the intermediate slopes of some of the mountains on the Lesser Antilles and in isolated locations and protected parks of the Greater Antilles. These areas are characterized by a profusion of species of plants, with a tree growth that typically produces interlocking canopies that shut out sunlight and keep the forest floor in semi-darkness. Long thick vines, called lianas, grow on large trees, as do many epiphytes such as orchids and Spanish moss. Similarly, such flowering plants as hibiscus, bougainvillea, heliconia, and philodendron grow profusely in these forests.

Proceeding upslope on the higher mountains above 3,000 feet (915 meters), the rain forest grades into a montane forest. Here the trees are smaller owing to the lower temperatures, which decline at a rate of about 3°F to 4°F (1.5°C to 2.2°C) per thousand feet. In some places, where clouds frequently cover the mountains because of its rain-producing effect, orchids, bromelaids, ferns, and mosses grow profusely, giving rise to cloud forests, also sometimes called elfin woodlands. At higher elevations in the montane forest, usually above 4,500 feet (1,372 meters), pines grow. Pines are found especially on the higher elevations of the Greater Antilles, and are less prevalent in the Lesser Antilles. Although pine barrens also exist on some of the flat coral islands—like the Bahamas, the Turks and Caicos Islands, and Cuba's Isle of Pines (now called the Isla de la Juventud)—these trees are a different variety that are stunted in the heights to which they grow. They are especially tolerant of sandy soils, as is the case with the well-known pine barrens of coastal New Jersey in the United States, but they do not have the value as lumber that the Caribbean pines that grow higher in the mountains do.

On the drier leeward sides of islands, in sheltered valleys, and on low-lying limestone islands the prevailing vegetation type is semi-deciduous woodland. Here the trees are short, have waxy leaves, and sometimes have thorns, all of which reduce moisture requirements. In the driest areas, drought-resistant plants like the cactus abound. If annual precipitation is much below 40 inches (1,015 millimeters) the tropical rate of potential evaporation exceeds rainfall and dry conditions prevail, enough for such xerophytic plants as cacti to grow.

Savanna grasslands, comprised of tall-growing, coarse natural grasses, are not very common in the Caribbean except on Cuba. Even in Cuba, most have been burned over to provide space to grow sugarcane. In most other parts of the world where they occur, savannas are caused by seasonal precipitation, but when they occur in the Caribbean, they appear to have been caused more by artificial, repeated burning of the natural semi-deciduous woodlands. One exception to this generalization is the island of Barbuda, in the Leeward Islands north of Antigua, where savannas have grown as a result of soil conditions. Here a layer of impervious clay a few inches beneath the topsoil inhibits the growth of roots, making it difficult for trees (with deeper root requirements) to grow.

Most land in the Caribbean has been cut over, so today most of it not in agricultural use is covered with secondary growth. This secondary growth illustrates the fragile nature of Caribbean ecosystems, because secondary forests are characterized by smaller trees and fewer species, so they have less value than the vegetation they replaced. Often what grows back, particularly in areas that used to be covered with semi-deciduous woodland, is scrub and thicket. This has been true especially in areas like Haiti and many of the smaller and lower-lying northern islands of the Lesser Antilles, such as the Virgin Islands, St. Martin, St. Barthélemy, St. Eustatius, Antigua, and the ABCs. Clear cutting the original vegetation also exposes the soil to erosion that may later narrow the biodiversity of plants that will grow there. Such destructive erosion has become a major problem throughout the Caribbean. More than half of Haiti has been devastated by erosion to the point that less than 20 percent of the country will now support any kind of forest growth. Estimates are that at least one fourth of Puerto Rico has similarly suffered severe soil losses. Another example is Barbados's Scotland District, which was cut over during the seventeenth century. After being planted in sugarcane, sheep were allowed to overgraze the secondary growth of grasses, causing massive gullies to develop. The result is a vast wasteland that occupies more than twenty percent of the island, an area that is likely never again to be of any significant agricultural use.

HISTORY

The Amerindian Population

The account here of the pre-Columbian Indian population is called the classical version of their history. Because there is little written record of these people, their origins and characteristics are still being debated by historians and anthropologists. For instance, a more recent view, called the new interpretation, suggests that the earliest Indians, the ancestors of the Ciboney, arrived in the islands from Central America or perhaps the Yucatán Peninsula of Mexico as early as 5000 B.C. A second invasion followed the first, but this time the invasion was from South America and may have occurred around 2000 B.C. According to this hypothesis it was the second stream from which both the Arawaks and Caribs evolved. The significance of this idea is that it suggests that the Arawaks and Caribs evolved from the same origin, rather than being representatives of different migration waves. Linguistic similarities and other common cultural traits are cited as evidence of the common origin of these two Indian groups. For a more detailed summary of this relatively new interpretation read Gary S. Elbow's "Migration

Versus Interaction: Reinterpreting West Indian Culture Origins," an unpublished manuscript at the Department of Geography, Texas Tech University, Lubbock, Texas 79409–1016.

Available evidence suggests that the Caribbean islands were occupied by at least three different Amerindian cultural waves prior to the arrival of Columbus in 1492. The earliest of these was the Ciboney culture, which may have migrated southward from Florida into the Bahamas and then throughout the Greater Antilles perhaps as early as 2000 B.C. Because they had almost disappeared by the time the Spanish arrived very little is known about them. There were probably a few thousand still living on the far western end of Cuba and in the southwestern corner of Haiti. They lived adjacent to the coasts and hunted, gathered, and fished but did not practice agriculture.

The arrival of the Arawaks from the northern coast of South America, around 300 B.C., represented the second wave of Amerindians in the Caribbean. Evidence suggests that these people traveled northward through the Lesser Antilles and into the Greater Antilles. By A.D. 1500 they occupied mainly the Greater Antilles. Apparently, as they swept westward through these larger islands they either absorbed or annihilated most of the earlier arriving Ciboneys.

Actually, the name Arawak was not used in the Caribbean but was given to these Indian people after the arrival of Europeans. The dialects they spoke were similar to that spoken by the Arawak Indians living in the Guianas of northern South America. Because of this linguistic similarity and common agricultural practices, later scholars lumped them together into one group. In the Caribbean they went by different names, depending on the islands on which they lived. For instance, those living in the Bahamas were called Lucayans, those in the Greater Antilles were the Tainan, and those in Trinidad, Tobago, and Barbados were the Ignerian. In several cases the names they gave their home islands later became the names of some of today's Caribbean countries. Thus, Colba became Cuba, Xaymaca became Jamaica, and Haiti became the name for today's Haiti.

The Arawaks were characterized by a higher level of technology than the Ciboneys. In addition to hunting, gathering, and fishing, they practiced agriculture and knew how to make pottery. In fact, the type of agriculture they developed is still practiced by many poor peasant farmers in the Caribbean, particularly on the mountain slopes of the poorer Lesser Antilles and throughout the Greater Antilles, especially in Haiti and the Dominican Republic. It was a farming strategy known as *conuco,* a variation of the shifting cultivation systems practiced throughout humid tropical areas in other parts of the world. The Arawaks first burned the plot of land they were going to use to remove weeds and also leaves on the trees so more sunlight could reach the ground. Then they piled the topsoil in round mounds that were sometimes knee-high and several feet (a meter or more) in diameter. These concentrations of more fertile soil were then enriched by ashes from the burning of the former forest cover. On these relatively fertile mounds—tropical areas are known for their heavily leached poor soils—the Arawaks concentrated their agricultural efforts, planting a wide variety of native crops including yuca (cassava or manioc), yams (sweet potato), arrowroot, peanuts, maize, beans, squash, cacao, various spices, cotton, tobacco, and numerous indigenous fruits such as the mamey and guava. Their diet was supplemented by fish, fowl, and other foods obtained from the nearby water and forests. When weeds encroached and soil fertility declined, the fields were abandoned, new land was cut over, and the process was restarted.

Although the conucos looked primitive and unorganized to Europeans because of the way crops were intergrown, in fact they represented a logical adjustment to the ecological conditions of the tropical Caribbean. Heaping the soil into mounds provided a loose, well-aerated soil; the earthen piles reduced sheet erosion by rainwater; the intercropping of different plants that grew to different heights provided ground cover as further protection against erosion; and the production of a variety of crops offered nutritious variety to the Arawak diet. Additionally, modern research has shown that this kind of companion planting effectively maximizes output of land.

The Caribs were the last Indian cultural group to arrive in the Caribbean. Like the Arawaks, they also migrated from northern

South America. However, they did not begin arriving until about A.D. 1000 and, as a consequence, they were found mainly in the Lesser Antilles. Evidence suggests that they were just beginning to encroach on Arawak turf in eastern Puerto Rico when the Spanish arrived. The name Caribbean was derived from these people.

The Caribs were more warlike, fewer in number, and less technologically advanced than the Arawaks, but they did practice some agriculture and, like both the Arawaks and Ciboneys, they also hunted, gathered, and fished. They replaced the Arawaks throughout most of the Lesser Antilles, except for Trinidad, Tobago, and Barbados. Because of their fierceness and reputed practice of eating the flesh of captured male enemies, whether for ritual purposes or merely for food, the word Carib came to mean cannibal.

Estimates of the number of Amerindians that occupied the Caribbean at the time Columbus arrived vary greatly, usually from 1 million to 10 million. Most scholars speculate that the correct figure was somewhere around 5 million. If so, what happened to these people is a tragedy with few precedents in human history and misery. The Ciboney had all but disappeared before A.D. 1500 and the Arawaks were to virtually vanish from the Caribbean within the next fifty years. The demise of the Arawaks was due to a number of factors such as war, importation of European diseases to which they had no immunity, destruction of their food supplies, and overwork in mines and as laborers on the Greater Antilles. In addition, thousands committed suicide, testimony to the cruel treatment they received from their Spanish conquerors.

The Caribs took longer to defeat as a result of their fiercer nature and because many retreated to the remote forested mountains of the Lesser Antilles. In addition, the Spanish did not consider the Lesser Antilles worth the effort it would take to conquer the Caribs because they had become more interested in the riches available on the Latin American mainland. It was the later-arriving northern Europeans who finally expended the effort to defeat the Caribs in the Lesser Antilles. The French subdued the Caribs in Grenada in 1651 but the British were not able to defeat them in St. Vincent until 1773. Today, there are probably no more than 3,000 recognizable Amerindians left in the Caribbean, and most of these are mixed bloods. The few remaining Caribs are found on the two islands of St. Vincent and Dominica. People with mixed Arawak ancestry are found on the Dutch ABC island of Aruba and very infrequently on the Greater Antilles.

Early Spanish Settlement

Christopher Columbus convinced the king and queen of Spain that the circumference of the earth was only 16,000 miles (26,800 kilometers) instead of the actual 25,000 miles (40,300 kilometers) that it is, thereby gaining their financial support for a voyage to the Far East. He planned to sail south from Spain to the Canary Islands where he could catch the northeast Trade Winds and then sail westward across the Atlantic, thereby finding a new and shorter route to the Orient. But instead, along the way he discovered a new hemisphere called the Americas. After a thirty-six-day voyage from the Canaries he landed on San Salvador Island in the Bahamas in October of 1492. From there he headed south and sailed along the northeastern coast of Cuba and then along the northern coast of Hispaniola before heading back to Spain. In 1493 he embarked on a second voyage during which he discovered most of the Leeward Islands, the Virgin Islands, Puerto Rico, Jamaica, and the southern coast of Cuba. Also during this three-year expedition he sailed twice (both coming and going) along the coast of Hispaniola. His third voyage in 1498 was his shortest. He sailed around the southern end and along the western coast of Trinidad before turning northwestward to sail along the south coast of Hispaniola and heading north through the Mona Passage between Puerto Rico and Hispaniola on his return to Spain. During his fourth and last voyage of 1502 to 1503 he sailed through the middle of the Lesser Antilles and, after passing along the southern coasts of Puerto Rico, Hispaniola, and Cuba, he turned westward to sail along the Caribbean coast of Central America before returning to Spain. Thus, in only twelve years Columbus had discovered most of the islands in the Caribbean, with the exception of most of the Windward Islands in the southern Lesser Antilles. It would be left to other less-known Spaniards to conquer these new lands and settle them: Ponce de León was largely responsible for settling Puerto Rico, as was Juan de Esquivel for Jamaica, Antonio de Berrio for Trinidad, and Diego Velásquez for Cuba.

For almost 150 years Spain reigned supreme in the Caribbean, although they were exposed to sporadic attacks by British, Dutch, and French pirates. But in truth, the Spanish soon had little interest in their Caribbean possessions, except for the few settlements they had established in the Greater Antilles, especially on Cuba and today's Dominican Republic. Within forty years the focus of interest had shifted first to Mexico and Central America and then a few years later to Peru and the former empire of the Incas. Still, the Spanish would leave an indelible imprint in the Caribbean through their construction of forts and cities, their architectural styles, and their place names, even on those islands where they would later lose control to other European powers.

Spanish construction of settlements in the Caribbean was determined by the Laws of the Indies, a codified body of laws promulgated by the Spanish government to guide the colonization process. Unless topography demanded otherwise, colonial towns were laid out with a well-defined gridiron street pattern, where streets intersected at right angles. In the center was a town plaza, around which activity of the settlement focused. The settlement's main church (usually the largest building by far) was located on or close to the plaza. Also, the most important government offices were located here, such as the mayor's office and the offices of the town council. Houses were usually built of stone or brick and covered with white stucco and barrel-tile roofs. Their walls were often made of thick masonry to keep out the tropical heat and to protect against hurricanes and they were aligned flush with the street, without front lawns. The better homes had interior patios, often ornate, with tile fountains and hanging plants, which took the place of the front lawns so common in North America. Most of the cooking also took place in a corner of the patio. Usually, the poor lived on the periphery of these settlements, while the middle class and wealthy lived near the town's center. Later, a few of these colonial towns grew into large, modern metropolitan cities and began to acquire some of the characteristics of North American and European cities. Nevertheless, traces of their original structures are still apparent in some areas even in these cities. The historically reconstructed center of San Juan, known as Old San Juan, is an outstanding example of a Spanish town.

Other European Competitors

No one seriously challenged Spain for control of the Caribbean until the seventeenth century. During the sixteenth century such activity as did occur was primarily in the form of piracy and privateers sailing under the flags or financial support of the British, French, and Dutch. The goal of these buccaneers was to plunder the main Spanish trading cities and intercept the wealthy convoys that were ferrying to Spain riches that had been stolen from the nearby Aztec and Inca empires. Favorite Caribbean targets were such cities as Havana, Santo Domingo, and San Juan, in addition to nearby mainland ports like Nombre de Dios, Cartagena, Santa Marta, Río de la Hacha, Puerto Cabello, and Isla Margarita. Privateers like the Englishmen Francis Drake and John Hawkins became famous as a result of their expeditions during the period between 1562 and 1596.

At least ten different countries would eventually play a role in settling the Caribbean—Spain, the United Kingdom, the Netherlands, France, Denmark, Sweden, the Knights of Malta, the United States, and the two German states of Brandenburg and Courland, in addition to independent privateers—but it was the British, Dutch, and French who competed most notably with the Spanish. The Dutch differed from the British and French in their goals in the West Indies because they were interested primarily in trading rather than in settling the islands. The few small islands they did settle were established primarily for trading purposes and none played a major role in the sugar industry that emerged during the 1700s. In addition, their period of important influence was shorter, lasting from 1570 until 1678. Still, during this period of a little more than a century their activities had a major and lasting influence on the Spanish. Their privateering activities damaged Spanish prestige and created a diversion so that the British and French could settle most of the Lesser Antilles. It was the Dutch who introduced the British and French to the plantation system of sugar production, owing to their earlier experiences in Brazil. They also provided through loans much of the capital for these early ventures. In addition, their shipping industry became the main provider of supplies and slaves for the other colonies located here. In fact, they were so successful with their trading activities that the British and French finally passed laws (called exclusives) that allowed their colonies to trade only with the mother country, much like the Spanish had done earlier. By 1678 the Dutch were eliminated from a significant role in Caribbean trade. As a consequence, they channelled their energies away from the Americas and into their East Indies empire. The only reminder of their earlier efforts in the Caribbean are their six small island possessions, including the three ABCs located adjacent to the northern coast of Venezuela and the three northern Lesser Antilles islands of St. Eustatius, Saba, and St. Maarten.

Unlike the Dutch, the British and French did engage vigorously in colonization. They chose, at first, to concentrate their efforts in those areas not given much attention by the Spanish, namely the Lesser Antilles. During the first half of the seventeenth century they occupied most of the Leeward and Windward Islands, with the exception of the few that had been settled by the Dutch and several that were still Carib strongholds, like St. Vincent and Dominica. The British occupied St. Kitts in 1623, Nevis in 1628, Barbados in 1625, and Antigua in 1632. Similarly, the French acquired both Guadeloupe and Martinique in 1635. During the latter half of the 1600s the British and French became more bold as they competed with the Spanish for possessions in the Greater Antilles and stepped up piracy against the Spanish colonies and their convoys traveling from the Caribbean to Spain. In 1655 the British captured Jamaica from the Spanish and in 1697 the French gained control of what is today Haiti (then called St. Domingue).

The eighteenth century was a period of frequent wars among the British, French, and Spanish. The greatest concern of each of the three countries was to maintain the territorial integrity of the home country, so their overseas colonies, especially those in the Caribbean, often became pawns in the diplomacy that followed each conflict. The result was a virtual state of sovereignty chaos in the Caribbean during the late 1600s and throughout the 1700s, as individual islands were swapped back and forth many times. St. Lucia changed hands fourteen times during the eighteenth century; while Tobago was passed back and forth between the British and French eight times between 1748 and 1814. Both Guadeloupe and Martinique had their ownership changed six times from 1761 until 1814. Evidence of this confusion is still found in the islands today in their place-names and vernacular speech patterns. For example, although St. Lucia, Dominica, St. Vincent, and Grenada became British possessions during the 1783 to 1803 period and remained so until their independence during the 1970s, many of their place-names are French and the local patois spoken informally is comprised mainly of a mixture of French and English words.

THE RISE AND NEAR DEMISE OF SUGARCANE

The first 150 years of settlement in the Caribbean were spent searching for a crop that could be profitably produced here and sold in Europe. In addition to food crops grown for subsistence, the colonists experimented with such cash crops as indigo, ginger, cotton, and tobacco. But the fact is that they did not grow enough food to feed even themselves, so from the start they were dependent upon importation of such food products as flour, rice, dried meat, and salt fish. Furthermore, by the middle 1600s they began to experience serious competition from the production of tobacco and, somewhat later, cotton in the United States.

Until 1640 tobacco was the most important cash crop grown in the islands. But then the Dutch introduced the British and French colonialists to the production of sugarcane. Within fifty years the agricultural foundation of the non-Spanish islands had changed radically. Sugar did not catch on as early in the Spanish colonies because of greater interests in exploiting their Latin American mainland colonies. The Hispanic islands were used mainly as way stations for supplying and protecting the Spanish fleets and for raising cattle to supply dried meat and leather for both Spain and the mainland. Although Spanish and Portuguese entrepreneurs from the Canary Islands and Madeira introduced sugar production in Hispaniola and Cuba by the 1500s, it did not become of major importance on these islands until 250 years later.

Sugarcane is ideally suited to the wet and dry tropical climate conditions of the Caribbean. It requires a year-round growing season of warm temperatures, and the omnipresence of high-intensity sunlight speeds its maturation. Furthermore, cane withstands droughts better than most other crops and benefits from a dry season that makes it easier to harvest. Also, harvesting can usually be finished before the period of greatest chance for hurricanes, during August and September. The rapidly increasing market for sugar in both Europe and the United States assured its early success. Because of the strict temperature requirements and the fact that most of the sugar was exported, plantations were usually located on level or gently rolling land along the coasts or in easily accessible valleys.

Sugar is not a crop that can be grown very profitably on small farms. It is best suited to large estates of several hundreds or even thousands of acres. This is because cane must be processed into sugar within forty-eight hours after it is cut and it is essential that a processing factory be located nearby. Thus, each estate had to have the necessary equipment and buildings to provide this essential function. This required large capital outlays, the costs of which had to be spread over a large farm. So, what had been largely a system of many small- to medium-sized farms reorganized into one of relatively larger estates. An elite element of society emerged, called the plantocracy, that controlled almost everything on the British and French islands. The stone ruins of their manor houses, factories, windmills, and equipment facilities still dot the landscape of almost all the islands today.

At first labor was provided by indentured immigrants from the British Isles and France because there was no longer an Amerindian population left to exploit. But soon it was clear that another source of workers was needed and this led to the importation of black slaves from western Africa. The infamous Trade Triangle developed, whereby cloth and manufactured goods were brought from Europe to Africa, slaves were imported to the Caribbean, and sugar, molasses, and rum were exported to Europe.

Although estimates vary, probably between 4 million and 5 million slaves were brought to the Caribbean, about the same number as were imported to Brazil and approximately ten times as many as were taken to the United States (to provide labor for its southern plantations). Most arrived during the eighteenth century, the day of the production of sugar in the Caribbean. The massive arrival of blacks drastically changed the racial composition of the populations in the British, French, and Dutch islands. Within a few years blacks were in the majority, a condition that accounts for the prevalence of an African-Caribbean majority today in all but the former Hispanic territories of Cuba, Puerto Rico, and the Dominican Republic. The Hispanic islands did not enter their sugar-plantation phase until the late 1700s. Thus, they did not import slaves for as long a period as was the case in the other islands and one consequence is that blacks comprise a smaller proportion of their populations. Probably about thirty percent of Cuba's population is comprised of blacks or mulattos, whereas in Puerto Rico the figure is between twenty and thirty percent. In the Dominican Republic the majority of the population is mulatto, but this is a result of an invasion from formerly French Haiti that lasted from 1822 until 1844, in addition to the plantation system.

During the latter half of the eighteenth century attitudes toward slavery in the British and French islands began to change. Certainly the American and French revolutions helped motivate this change, but so did the inhumane nature of the system itself promote its near demise. First the British abolished the slave trade in 1807, followed by the French in 1818, and the Spanish in 1820. The British emancipated their slaves in 1834, but this was followed by an apprenticeship period of four years, until 1838, when the former slaves were required to remain and work on the plantations as wage laborers. The French freed their slaves in 1848, whereas Puerto Rico and Cuba released theirs in 1873 and 1886, respectively.

Emancipation created a labor shortage in the Caribbean because many of the freed slaves left their former plantations, most to work on their own small peasant farms. As a result, there was a scramble to find another labor source. Indentured workers were once again enticed to move to the West Indies. Although they came from a variety of sources, the largest proportion came from British India, and secondarily from China. Between 1835 and 1917 almost 700,000 such workers were imported to work in West Indian cane fields. The usual duration of indenture was five years, after which most workers left the estates. As a result, indentured workers did not provide a long-term solution to the labor problem. But they did arrive in large enough numbers to significantly affect the populations of such islands as Trinidad, Jamaica, Guadeloupe, Grenada, and St. Martin. In fact, in Trinidad, persons of southern Asian descent represent about the same percentage of the population as African descendants. On this island they are still associated with the sugar industry and are more than proportionately found in rural areas. The Asian Indians have had a significant effect on local politics and the landscape, the latter especially through construction of their Hindu temples and Moslem mosques. However, on the rest of the islands they represent a much smaller but still visible minority and live primarily in the cities where they are often merchants.

By the middle nineteenth century the Caribbean sugar industry, with the exception of the Spanish colonies, was clearly in decline. It started to decline near the end of the 1800s on the Spanish islands because of the later dates of slave emancipation there. A number of reasons account for the deterioration of the Caribbean sugar industry. With emancipation, the cost of labor significantly increased. Soils in some areas were becoming badly eroded and otherwise depleted of their fertility. Estate supplies imported from Europe had increased in price. The price supports that the British and French used to provide to the West Indies were discontinued and the formerly captive markets with the homelands were opened to outside competition by the middle 1800s. Because their markets used to be protected, there was little incentive for West Indian producers to modernize their facilities and operations. As a consequence, when this protection was removed the inefficiency of the British and French operations made it difficult to compete with new competitors, such as Cuba, Mauritius, and a resurging Brazil. In addition, sugar-beet production in Europe, which was usually heavily protected and subsidized by local governments, provided another source of competition. In the United States it was discovered during the 1970s that sugar can be derived from corn syrup and today this source has captured almost forty percent of the U.S. sugar market. Finally, as Americans have become more health and weight conscious there has been a shift toward use of less caloric sugar substitutes, such as saccharin and aspartame.

Today the sugar industry has all but disappeared from most of the Caribbean islands and is only of significance on a few islands such as Cuba, St. Kitts, Barbados, Trinidad, Guadeloupe, the Dominican Republic, and Jamaica. In fact, on all of these islands, except Cuba, there also has been a significant decline in land used for growing sugar. For example, by the middle 1980s Jamaica was producing only about half as much sugar as it was during the 1960s. During the early 1970s almost one third of all Barbados's exports by value were accounted for by sugar, but by the middle 1980s the figure had declined to about two percent. In 1982, for the first time in more than 100 years, Puerto Rico had to import sugar from the U.S. mainland to meet its own needs. The only way the sugar industry has been able to survive at all in the Caribbean is due to renewed favorable trade agreements with other countries, especially those included in the European Economic Community and North America. Similarly, the Soviet Union and Eastern European countries used to buy most of Cuba's sugar production at above-market prices.

The Drive to Independence

The colonial policies originally developed by the various European countries in the Caribbean were based on a system of exploitation of the people and resources. Enormous wealth was generated in the islands for the home countries. During the seventeenth century these were considered the most valuable colonies in the British and French empires. The British West Indies were often referred to as the jewels of the British Crown. In 1763, at the end of the Seven Years' War between the British and French, the French (who lost) decided that they would rather give up their part of Canada rather than their West Indian possessions.

The results of the American and French revolutions during the later 1700s kindled hope of liberation in the exploited British and French colonies. The independence of Mexico in 1821 and the Central American countries in 1838 did the same for the Spanish islands. Haiti was the first Caribbean state to break free, when it became independent from France as a result of a slave revolt that lasted from the 1790s until 1804. The Spanish colonies followed next, with independence coming first to the Dominican Republic in 1834. Cuba and Puerto Rico were freed from Spanish rule in 1898 as a result of the Spanish-American War. Cuba became independent in 1902, but Puerto Rico remains affiliated with the United States through its commonwealth status and internal self-government. With the exception of Haiti, the former French colonies have remained associated with France and have been elevated to départment status, as the Hawaiian Islands have become the United States' fiftieth state. The six Dutch islands similarly remain associated with the Netherlands, although Aruba is tentatively scheduled for independence in 1996. In 1917 the United States purchased the U.S. Virgin Islands from Denmark, and since then these islands have been affiliated with the United States in a way similar to that of Puerto Rico.

The current and former British West Indies have had the most complicated road to independence of the Caribbean territories. Some, like Anguilla, the British Virgin Islands, Cayman Islands, Montserrat, and the Turks and Caicos Islands remain possessions of the United Kingdom with internal self-governments. Others, like most of the former British Lesser Antilles and the Bahamas, have become independent. The first to become independent were Jamaica and Trinidad–Tobago in 1962. Barbados became independent in 1966 and the rest achieved independence during the 1970s and early 1980s, with St. Kitts and Nevis being the most recent to become independent in 1983.

In 1959 Cuba became the only country in the Caribbean to experience a Marxist revolution. Grenada experimented with Marxist–Leninist ideals from 1979 to 1983, but internal problems (viewed as a threat to its Windward Islands neighbors) prompted an invasion by a coalition force comprised of U.S. and eastern Caribbean troops. Today the Caribbean contains twenty-four different political entities ranging from independent countries to overseas possessions of several types.

ECONOMIC GEOGRAPHY

In terms of their economic well-being, the Caribbean islands exhibit a great deal of diversity (see the Appendix). Haiti is by far the poorest, with a level of poverty that approaches that of the poorest African and southern Asian countries. On the other hand, the wealthier islands, like the Bahamas, the U.S. Virgin Islands, the Cayman Islands, the French West Indies, and the Dutch ABCs, compare favorably to some of the poorer southern European countries, for instance Portugal and Greece. Still, there is no doubt that, despite this variation, the Caribbean is generally a Third World (poor) region. However, with the exception of Haiti, this is not as poor an area as most other less developed regions. It is much better off than most of Africa and southern or southeastern Asia. Furthermore, its standard of living is higher than that of Central America and somewhat above that of most of South America. On the other hand, the Caribbean is clearly poorer than most of the Western and Northern European countries, Japan, Canada, and the United States. For instance, the per capita gross national product of the Bahamas, one of the wealthiest Caribbean countries, is only fifty-four percent that of the United States.

Agriculture

The decline of the sugar industry has been so far-reaching that it is only a significant agricultural pursuit on seven of the islands in the Caribbean. But agriculture has not disappeared from most of the islands where sugar is no longer important, because other crops have also played a significant role in the farming pursuits of the Caribbean.

Spanish missionaries introduced bananas during the early 1500s, but they did not become important as a major crop until the 1880s. By the beginning of the twentieth century bananas had become popular in Europe and the United States, and Jamaica was the world's leading producer. There are two general types of bananas that are important in the Caribbean. One is the sweet dessert variety with which most Americans and Europeans are familiar. It is the type that is largely exported. The second is the cooking variety, called plantain. The latter type is domestically used in the West Indies and is not exported in nearly as large quantities as the dessert type. It provides starch in the Caribbean diet and is boiled, fried, or roasted. Sweet bananas, the kind exported, are exacting in their requirements. They are most easily grown on fertile alluvial or volcanic soils and on the wetter windward sides of the islands, where annual rainfall exceeds ninety inches (2,286 millimeters). Most often they are cultivated on large plantations, where they are grown under carefully supervised conditions. Plantains tolerate less rainfall and are more frequently grown on smaller peasant farms on poorer soils both on coastal lowlands and in the foothills of the mountains. On several of the Windward Islands of the Lesser Antilles sweet bananas have replaced sugarcane as the main agricultural crop. Examples of this are St. Lucia, St. Vincent, Grenada, Martinique, the Basse-Terre side of Guadeloupe, and Dominica. Bananas also figure as an important secondary product in the Dominican Republic and Jamaica. But the Caribbean is no longer a leading producer of bananas. Far more are grown in Central America, South America, India, and several of the southeastern Asian countries.

Tobacco was one of the earliest cash crops grown in the West Indies. In fact, it was used by the pre-Columbian Indians and may even have originated in the Caribbean, or possibly in South America. Today, however, it is of greatest significance in Cuba and the Dominican Republic. A little is also grown, mainly for domestic use, in Jamaica and Puerto Rico.

Coffee was at one time important as a mountain-grown cash crop on all the Greater Antilles and its cultivation eventually extended as far south as Trinidad. But production in the Caribbean could not compete with coffee production in some of the Central American countries and particularly Columbia, Brazil, and Venezuela in South America. Some coffee is still grown for export on mountain slopes in the Dominican Republic, Jamaica, and Haiti. A small amount, mainly for domestic use, is also grown in Puerto Rico.

Marijuana was introduced to the Caribbean during the middle 1800s by Asian Indians who brought it with them as a holy plant when they arrived as indentured laborers in the British islands. It rapidly grew in importance during the 1970s and 1980s, to the point that it is now a more than billion-dollar industry. It is most often consumed through cigarettes or cigars (called spliffs) and had become associated with the region's Rastafarians, a religious and cultural sect. The island that is most associated with its growth is Jamaica, but it is grown in remote mountainous areas throughout the Caribbean, despite the fact that it is universally illegal. (Tourists are warned that it can be hazardous to hike in remote areas unknown to them. It is best to hire a guide when hiking in areas with poorly marked trails.)

Cacao, used to make chocolate, originated on the adjacent Central American mainland, maybe in southern Mexico. It is another crop that used to be widely grown throughout the Caribbean, but is not so widespread anymore because it is so vulnerable to the high winds associated with hurricanes. It is still mildly important today only in the Dominican Republic, Trinidad, and several of the Windward Islands. The world's leading region of production now is western Africa, especially Ivory Coast and Ghana.

Other significant commercial crops grown both for export and domestic use in the Caribbean today are pineapple, coconut, and citrus. Pineapple and coconut are grown almost exclusively on the coastal lowlands and citrus trees are raised sometimes on the lowlands (as in Cuba) and sometimes in the low mountains (as in Jamaica and Puerto Rico). But these three crops are grown on almost all the islands in small quantities. The Caribbean is not one of the world's leading exporters of any of these three crops, although Cuba has provided much of the citrus consumed in Eastern Europe and the former Soviet Union.

Peasant farming is still common in the Caribbean. These are small-scale operations, usually involving less than ten acres (four hectares) of land. The origins of many peasant farms can be traced to the emancipation of slaves, when they left the plantations to strike out on their own in the land that was left over. Usually, the land is of poor quality, with thin soils, and often located on lower slopes of mountains. Although this land is not suitable for modern mechanized agriculture, it can be used for small-scale operations that use labor-intensive and non-mechanized techniques. Some peasant farmers work as associate producers: They sign a contract with a commercial corporation to produce bananas, sugar, and sometimes other crops such as cacao, citrus, and spices like nutmeg and mace. Others work as semi-subsistence producers to meet the needs of their own families and sell whatever surplus they have at the markets in the nearest town. Often the men clear the land in a fashion similar to the type of conuco agriculture practiced by the Arawaks before Columbus arrived. Women sometimes help plant, tend, and harvest the crops and it is usually the wife who takes the surplus to market for sale. Peasant farming is usually a part-time occupation in the Caribbean, as most farmers hold other jobs during at least part of the year. Often they work in the tourist industry or manufacturing, or help cut cane on the sugar plantations during its harvest season. Today, approximately forty percent of all people living in the Caribbean reside in rural areas, indicating that agriculture is still important here.

Despite the fact that farming is still widely practiced in the Caribbean, it is also true that not nearly enough food is produced to meet domestic needs. At least fifty percent of the food consumed in the area is imported from other countries, usually located outside the region. In some cases, such as Antigua, Puerto Rico, Barbados, Trinidad, and the Bahamas, the amount imported approaches eighty percent. At least four factors account for the Caribbean not being able to feed itself. One is that population densities are so high on many of the islands that they could feed themselves only by developing an intensive form of farming similar to that practiced in some parts of eastern and southeastern Asia. Second, much of the best agricultural land is used for cash crops. Food crops are usually grown on the peasant farms, that have steeper slopes, less fertile soils, and generally poorer growing conditions. Third, tastes in the Caribbean have changed during the past thirty years. Now many West Indians prefer to eat food imported from the United States and Europe, such as wheat-flour bread, noodles, potatoes, beer, and whiskey, rather than eating and drinking locally produced cassava, sweet potatoes, breadfruit, and rum. A fourth factor is the success of tourism. Tourists consume much of the food that might otherwise be eaten by local inhabitants. As a consequence of these factors, food prices are fifty to 100 percent higher in the Caribbean than for comparable items in the United States and Europe, a frequent complaint of tourists.

Mining

Generally speaking, the Caribbean is not blessed with mineral wealth, with three exceptions: Cuba, Jamaica, and Trinidad. Cuba has the greatest variety of metallic minerals, most of which are located on its mountainous eastern end. It exports some manganese, cobalt, nickel, chrome, and iron to Eastern Europe and the former Soviet Union from this region, plus it has small copper reserves east of Havana near Pinar del Rio. But, exports of these products do not figure as proportionately important in Cuba's economy as minerals do in Jamaica and Trinidad.

Jamaica has large reserves of bauxite, the ore from which aluminum is made. Jamaica is the world's third leading producer of this ore, most of which is found in the valleys among the limestone mountains in the center of the island. Because of the weight-losses involved, some of the bauxite is processed into intermediate grade alumina before being shipped elsewhere for final processing into aluminum. No aluminum is manufactured in Jamaica because of the high energy requirements of this final processing stage, and as a result most of the alumina is shipped to either the United States or Canada and then refined into aluminum. The exportation of both bauxite and alumina provides for almost sixty percent of Jamaica's exports by value, thus representing the single most important element in the island's economy.

Energy Production

The Caribbean islands have abundant sunlight and wind. Wind was a significant source of windmill-generated power for many of the small sugar factories that occupied the West Indies during the sixteenth and seventeenth centuries, and sunlight and wind may once again be major sources of energy.

Trinidad is the only island in the Caribbean to have significant reserves of oil and natural gas. Trinidad's oil and gas industry provides for almost eighty percent of the value of its exports and supplies jobs for approximately 20,000 people. The production of oil and gas, and its refinement into various petroleum products, is by far the island's most important industry.

Other islands have refineries but none produce their own oil, with the partial exception of Barbados, which has small reserves but not nearly enough to satisfy even its own needs. In fact, the Caribbean is badly deficient in traditional energy reserves, a dilemma that complicates its problems of economic development as petroleum prices continue to rise. Still, the refining of imported oil is big business on some of the islands. About one sixth of the oil consumed in the United States is refined in the Caribbean.

The Caribbean has a number of oil-transshipment terminals in addition to its refineries. Approximately fifty percent of the oil imported to the United States passes through Caribbean shipping lanes. The reason for this is that the largest oil tankers that transport oil from places like Venezuela, Nigeria, and the Middle East are not allowed access to some ports in the eastern United States because of their size and because of environmental concern. The supertankers unload their crude oil at Caribbean ports, where it can be transferred to smaller ships to gain access to the lucrative markets on the U.S. eastern seaboard. Most West Indian ports that serve as transshipment terminals also take advantage of the unloading of the oil to refine some of it, thereby benefiting from the profits achieved by the value added in this processing stage. There are, however, a few exceptions, where only transshipment occurs without refining, such as Bonaire, St. Eustatius, and St. Lucia.

Manufacturing

Until recently, most of the manufacturing in the Caribbean was for the local market and mainly involved food processing, the making of clothing, or the manufacturing of sugar and rum. Since the 1970s, however, most island governments have made major efforts to expand these activities as a way of earning additional income and providing new jobs. These efforts have been keyed to the concepts of import substitution and industrialization by invitation.

The phrase import substitution refers to a policy of trying to produce goods that were formerly imported. The advantage is that goods locally produced will benefit from the value added and jobs created during the manufacturing process. Thus, many island governments have imposed tariffs on the importation of foreign goods to encourage the development of local industry. Examples of the types of manufacturing that have benefited from this protection and encouragement are soft drinks, locally produced garments, locally manufactured food products, pharmaceuticals, and alcoholic beverages.

Industrialization by invitation is a strategy aimed at attracting foreign capital for investment in local industry. Among the incentives used to attract this investment capital are low-cost labor, factory accommodations constructed by the local government, low taxes or complete tax abatement for a number of years, government-sponsored training programs, political stability, and proximity to the large North American market. Many of the factories attracted by this strategy (called offshore industries) are located in free trade zones where no import duties are charged on the imported ingredients. The goods produced are almost always exported. Often these establishments are also referred to as finishing touch industries or screwdriver industries because almost all the components are imported and only the final assembling takes place on the islands.

Companies assembling goods for export to the United States benefit from special U.S. tariffs that either reduce or waive import duties in these products. When duties are imposed, they usually are assessed only on the value added to the products by the Caribbean operations. Among the types of manufacturing that have benefited from these arrangements is the garment industry. U.S. firms, wanting to escape high-cost unionized labor, have established factories in the Dominican Republic, Haiti, Jamaica, and Barbados. The workers hired are predominantly female and they typically earn only between $30 to $70 per week. There are perhaps as many as 800 such factories throughout the Caribbean employing more than 20,000 workers. A significant portion of the less-expensive women's undergarments and children's clothes sold in the United States are now made in these island factories.

Tourism

A number of factors have conspired to make tourism a big industry in the Caribbean. The natural tropical island environment with some of the world's most beautiful beaches has produced an often dramatic and idyllic setting for a vacation. The varied history has created a diverse cultural setting in which it is easy to visit English-, French-, Dutch-, and Spanish-speaking islands during one trip. The nearness of the Caribbean to the wealthy North American market has given it an advantage over more distant tropical island environments.

There was a small but significant Caribbean tourist industry, with a primary concentration in Cuba and a much smaller secondary core in Jamaica, prior to World War II. But not until the 1950s when regular and inexpensive air service was available did the industry become significant. The Puerto Rican government made major efforts to build tourism during the late 1940s and early 1950s, but the demise of tourism in Cuba that resulted from the communist revolution in 1959 was the major reason for the success of Puerto Rico's tourism during the early 1960s. This success spurred similar government promotion and development in the U.S. Virgin Islands and the Bahamas during the 1960s and 1970s. Their success, in turn, motivated other islands to promote tourism. Today, tourism accounts for at least ten percent of the gross national product (GNP) of the Caribbean. This sector of the economy also generates about twenty percent of the region's export earnings, making tourism the number-one earner of foreign currency.

Of course, the level of importance of tourism differs throughout the Caribbean. For instance, tourism generates between seventy and eighty percent of the GNP of the Bahamas and the Cayman Islands. It accounts for between thirty-five and fifty percent of the GNP in the U.S. Virgin Islands, Antigua, and Barbados. But it provides less than ten percent of the GNP in some of the Lesser Antilles like Dominica, St. Vincent, St. Lucia, Martinique, Guadeloupe, and Trinidad and Tobago and in the Greater Antilles countries of Haiti and the Dominican Republic.

Although tourism is the leading earner of foreign currency for the Caribbean, the industry here is not big by world standards. Only about three percent of the world's tourists and about five percent of its earnings go to this region. The number of tourists, and the dollars they generate, are much smaller than in Europe and North America. However, compared with other Third World countries, tourism is well developed in the Caribbean. By the end of 1991 almost 10 million tourists visited the Caribbean.

There are two sides to the Caribbean tourist industry. One is overnight stayers and the other is cruise-ship passengers. Generally, the overnight stayers spend more money on an island, but both are important generators of income. In terms of arrivals of overnight stayers Puerto Rico and the Bahamas are the leaders, together accounting for almost forty percent of all those visiting the Caribbean. In terms of earnings, the U.S. Virgin Islands joins the top of the list with Puerto Rico and the Bahamas. The reason that the U.S. Virgin Islands ranks higher in terms of expenditures than it does with respect to arrivals has to do with its favorable import-duty position with the United States. Because it is a possession of the United States, U.S. residents are allowed to bring back twice the value of duty-free goods that they can from the rest of the Caribbean islands.

The cruise-ship industry is even more dominated by Puerto Rico, the Bahamas, and the U.S. Virgin islands than the overnight stayer component. Almost fifty-five percent of all cruise-ship passengers traveled to these three territories. It is clear that both the number of overnight stayers and cruise passengers is mainly affected by two factors, nearness to the United States and efforts the island governments have invested in advertising and developing their tourist industries.

Almost two thirds of the tourists who visit the Caribbean live in the United States, another eight percent live in Canada, and an additional ten percent in Europe. Although the number of inter-island tourists visiting within the Caribbean has increased, as a proportion of the total they have decreased since 1970. With the rise of a substantial middle class in the region, more local people are traveling to foreign areas outside the region for their vacations, rather than traveling within it from one island to another.

Contrary to popular belief, Caribbean tourism is not characterized by one season but rather three. The time of greatest travel is during the cold winter months of February and March in North America and Europe. The second highest season is the month of December, during Christmas vacation. A third, more minor, season occurs during the late summer months of July and August. Prices in hotels, on cruise ships, and airlines reflect this seasonality.

Offshore Banking

The phrase offshore banking refers to financial operations conducted by foreign banks that have branches in countries like
those in the Caribbean. The attraction of conducting business in
Caribbean countries is that the countries provide a number of
advantages over keeping all assets in the home country, such as
the United States, Canada, or any Western European country.
Absence of taxes on income, profits, dividends, and capital
gains are one attraction. The Caribbean countries involved also
have modern and reliable communications systems. Accounts
are kept secret, causing some to speculate that these operations
have become "laundry" facilities for illegal activities, such as
the drug industry. The political stability of the Caribbean countries is another advantage over banking in less stable Third
World countries.

The reason that a country would like to attract foreign banking
is that legal fees and licenses are charged to the banks, but these
expenses are usually less than those that would incur if the activity
took place in the home country. Such fees add valuable foreign
currency to the island's economy where the activity is taking
place.

Offshore banking is a relatively new industry in the Caribbean.
The Netherlands Antilles, especially Curaçao, was the first to
develop this industry, but recently it has been curtailed there. Now
the leading centers for offshore banking in the Caribbean are the
Bahamas, the Cayman Islands, Antigua, the Turks and Caicos
Islands, Montserrat, and St. Vincent. In the Bahamas and the
Cayman Islands, offshore banking is the second leading industry,
behind tourism, providing for between fifteen and twenty percent
of each country's GNP.

POPULATION PATTERNS

Population Distribution and Density

In the Caribbean, island size and population relate clearly and directly: The larger the island, the larger the population. The Greater Antilles collectively contain eighty-nine percent of the region's population, a figure identical to their share of the total area. An exception to this rule is the Bahamas. It is also clear that more than four of every five residents of the Caribbean are culturally of Latin descent and Catholic in their religious preference. Almost sixty-three percent live in the former Spanish territories of Cuba, the Dominican Republic, and Puerto Rico. Another twenty percent live in the former French colonies of Guadeloupe, Haiti, and Martinique.

The growth of population in the Caribbean over the 500-year period of European influence has generated extremely high population densities. In fact, population pressure is nothing new in the Caribbean. The population density of the islands has long been much higher than that of the United States. It has been estimated that the West Indies (not including the Greater Antilles) had person/land densities that exceeded 100 people per square mile even before the middle 1800s, higher than the United States' density today. The reason for the high Caribbean densities can be traced to the labor demands of sugar production, especially on the British, French, and Dutch islands.

In 1960 the islands in the Caribbean had approximately 17 million inhabitants. By 1991 this figure had doubled to 34 million, for an annual (geometric) growth rate of 2.2 percent. This was a rate of growth that is about twice that of the United States for the same period. Since the 1960s, island birth rates have declined, but at the same time death rates have also decreased. As a consequence, the rate of natural increase (the birth rate minus the death rate) has declined only modestly to about 1.8 percent. Were it not for emigration, a growth rate this high would have stifled any attempts at economic development because the population would have doubled every nineteen years.

With a population of 34 million and an area of approximately 92,000 square miles (35,510 square kilometers), the Caribbean has a person/land density of 370 people per square mile, a figure more than five times that of the United States. But this figure is virtually meaningless because the islands in this region vary tremendously. The country with the lowest person/land density is the Bahamas (47 people per square mile or 18 people per square kilometer) and the highest ratio is found in Barbados (1,548 people per square mile or 598 people per square kilometer). The average person/land density of all the islands, unweighted by population, is 584 people per square mile (225 people per square kilometer), a figure over eight times that of the United States. The density figure of 584 persons per square mile is derived by taking the person/land densities for all the Caribbean countries listed in the Appendix and averaging them. Thus, it provides an average for all the islands but disregards the islands' vastly different population sizes. The density value of 370 people per square mile (143 per square kilometer) is a population-weighted figure.

But the person/land figures do not consider the fact that only about one fourth of the land in the islands is available to feed their populations, due to steep mountain slopes, erosion, poor soils, insufficient rainfall, swamp conditions, and competition from urban, industrial, and other uses. A better indication of population pressure is to consider only the land being used for agriculture (not including animal grazing), a measure called the physiological density. Physiological densities range from 831 people per square mile (321 people per square kilometer) of cultivated land in Cuba to 8,275 in the Netherlands Antilles. The average for all the islands (unweighted by population) is 2,677. No wonder the Caribbean cannot feed itself and must import more than fifty percent of its food! No other major region in the world has such heavy population pressure.

In fact, even physiological densities do not adequately capture the concept of well-being of the Caribbean's population because more than half its residents live in urban areas, where they derive their support from nonagricultural activities. The figures in the Appendix clearly show that there is no logical relationship between physiological densities and per capita GNP. For instance, both the Dominican Republic and Haiti have lower than average physiological densities and the lowest per capita GNPs. Conversely, both the Netherlands Antilles and Puerto Rico have among the highest physiological densities and higher than average per capita GNPs.

Migrations

From its very beginning, the Caribbean population has been migratory. Even the aboriginal Indians who occupied the islands before the arrival of Europeans came from someplace else. Before World War II the major migrations to and from the Caribbean included:

- □ immigration of Amerindians from Florida and northern South America;
- □ immigration of Europeans from the main colonial countries such as the United Kingdom, France, the Netherlands, and Spain;
- □ immigration of a small but significant number of Jews from Brazil to Curaçao and Jamaica;
- □ immigration of blacks from western Africa imported during the slave trade;
- □ immigration of indentured laborers imported from 1834 until 1917 to the British, French, and Dutch possessions;
- □ emigration of West Indian blacks to work on the Panama Canal during the 1880s for the French and again between 1904 and 1914 for the United States;
- □ emigration of West Indians primarily from Barbados and Jamaica to work on the sugar and banana plantations along the Caribbean coast of Central America;
- □ emigration from the West Indies to the oil fields of the Maracaibo Basin in Venezuela and to the refineries on Aruba and Curaçao between 1900 and 1930;

- immigration mainly from Spain to work in Cuba's rapidly growing sugar industry between 1898 until the Great Depression of the early 1930s;
- emigration to the United States from the 1850s until 1924 when the National Origins Immigration Act was passed by Congress; and
- intra-Caribbean migration from the more densely populated Lesser Antilles to Trinidad, Cuba, and the Dominican Republic to serve as laborers on sugar plantations.

After World War II heavy emigration resumed from the Caribbean islands. Most of these emigrants went to either the European home country—the United Kingdom, France, the Netherlands—or to the United States and Canada. For instance, during the 1950s the movement from the British West Indies to the British Isles became so great that in 1962 the British government passed its Commonwealth Immigration Act, which greatly restricted this in-flow. Since then, most of the people leaving the former British islands have gone to the United States or Canada. By the middle 1980s more than 7 million emigrants from the Caribbean lived either in Europe or North America, representing almost one fourth of the population still residing in the islands. About eighty percent of the island emigrants lived in the United States, another fifteen percent in Europe, and five percent in Canada. Of those who emigrated to Europe, almost fifty-six percent went to Great Britain, twenty-six percent to France, and eighteen percent to the Netherlands.

Despite the concentrated movement to Europe, it was the United States that received by far the greatest number of island emigrants. By 1985, almost 6 million people from the Caribbean lived on the United States mainland. Although this represented less than three percent of its total population, their presence was more noticeable because they tended to favor several metropolitan areas. For instance, the 2.6 million Puerto Ricans clustered in the northeastern states, especially in the metropolitan area of New York City. The 1 million Cubans concentrated in metropolitan Miami, with a secondary cluster in New York City and adjacent New Jersey. The 800,000 emigrants from the Dominican Republic preferred New York City, as did the 750,000 Haitians. A secondary cluster of

Haitians occurred in metropolitan Miami. Clearly, immigrants from the islands favored the metropolitan areas of New York City and Miami, although thousands also went to other cities in the United States. In addition to the migration to the mainland, there was a notable immigration to the U.S. Virgin Islands from Puerto Rico, the British Virgin Islands, and the nearby less prosperous Leeward Islands between World War II and the late 1970s, but most of this has now been stopped.

Although the emigrants from the Caribbean represent only a small proportion of the populations of the European and North American countries they moved to, they represent a very significant percentage of the island they moved from. If these people were to suddenly return to the Caribbean islands, the result would be nothing short of catastrophic. Today, almost forty-five percent of all Puerto Ricans live in the United States, and if they were to return to their island homeland its population would increase by almost eighty percent. Similarly, the Barbadian– and Jamaican–Americans represent one fourth and one fifth of the populations living in Barbados and Jamaica, respectively.

Because Caribbean immigrants tend to concentrate in a few cities in the United States, the city with the second largest number of an island's natives is often located outside the Caribbean. For example, metropolitan Miami has more Cubans living in it than does Santiago, Cuba's second largest city. Similarly, greater New York City has become the city with the second largest number of Caribbean nationals for Puerto Rico, the Dominican Republic, Haiti, Jamaica, Barbados, and Trinidad. Montreal is third for Haiti; Paris is the second for both Martinique and Guadeloupe; London is the second for several of the former British Lesser Antilles; and Amsterdam is the second for some of the Netherlands Antilles.

As important as emigration has been to the Caribbean in serving as an escape valve for population growth, it should not be viewed as a solution for this region's building population pressure. At best it should be regarded as a factor that temporarily slowed what would otherwise have been ruinous population growth on the islands. Some countries that once received significant numbers of immigrants from the Caribbean are now tightening their immigration requirements. Great Britain's Commonwealth Immigration Act was passed in 1962, and Canada tightened its immigration policy in 1972, as did the United States in 1986. Yet, today the United States continues to "welcome" Caribbean immigrants despite recent legal changes in entry requirements and procedures.

At best, emigration has provided more time for these islands to solve their problems of high natural growth rates. Still, it is necessary for these countries to reduce their fertility rates below their current moderately high levels. Happily, this appears to be happening: Evidence suggests that fertility is declining generally throughout the Caribbean. Even within the Caribbean serious efforts have been made to reduce inter-island moves. For instance, Haitian and Jamaican migrations to Cuba have been stopped. The Netherlands Antilles have halted the influx from the British and French islands. Jamaica and Barbados forbid the entry of unskilled laborers. Trinidad, much of whose black population initially came from the Windward and Leeward islands, now has legal barriers against further unrestricted entry. The only island immigrants who are not discouraged are the skilled and professional groups, and this kind of brain drain is precisely what the islands do not need.

URBANIZATION

People have been migrating from rural to urban areas in the Caribbean since the early 1900s, but this movement has become especially significant since the Second World War. Today it is one of the most important migrations affecting this region, as the island populations are rapidly changing from a predominantly rural and agricultural orientation to an urban focus. In 1960 fewer than forty percent of all the Caribbean population lived in cities. By 1991 this proportion had increased to almost sixty percent, and by the year 2000 it is expected to increase to sixty-five percent. Already some of the islands have levels of urbanization equal to that of the United States (seventy-four percent), but others are much lower. As with everything else, the Caribbean shows great variation. The highest level of urbanization is found on Guadeloupe, where ninety percent of its inhabitants live in urban places. The Bahamas (seventy-five percent), Dominica (seventy-three percent), the Netherlands Antilles (eighty-two percent), and Puerto Rico (seventy-two percent) also have higher than average percentages of urban population. The lowest proportions tend to be found on the poorer islands; Haiti has the lowest percentage of its population living in cities (twenty-eight percent).

Generally speaking, people have been moving from the countryside into the cities for two main reasons, the decline of economic opportunities in the rural environments and the perceived increase of such opportunities in the cities. As a result, the rural populations are growing at average rates of less than one percent throughout most of the Caribbean, whereas the growth of the cities exceeds three percent. In fact, on some islands the rural population is declining, as it has been in Puerto Rico since 1960. In 1950, only seven cities had a population of greater than 100,000 in the Caribbean and only Havana had more than a million residents. Now at least twenty-five cities have populations greater than 100,000, and five metropolitan areas have populations exceeding 1 million (Havana, Santo Domingo, San Juan, Port-au-Prince, and Kingston).

Much of the growth in the Caribbean is taking place in each island's largest city. Usually, this one city, with its surrounding metropolitan area, dominates the urban structure of each island. When the largest city is more than two or three times as large as its next competitor it is called a primate city. Primate cities are primary in almost every respect, dominating the island's economy, culture, politics, and social scene. With the exception of Guadeloupe, whose capital is the small city of Basse-Terre, each island's primate city is also its capital.

The annual growth rates of the largest metropolitan areas often exceed the three percent annual average for all Caribbean urban areas, but even a rate of growth of three percent means that a city's population will double in twenty-three years if it continues into the future. It also means that its population, if unchecked, would double three times in about seventy years, so it would be eight times as large as it is now. Clearly, such explosive urban growth creates tremendous problems in providing housing, jobs, and social services for the new inhabitants, about one fourth of whom are immigrants and the remaining three quarters are the product of the excess of births over deaths.

The problem of housing has become particularly acute in the primate cities of the Caribbean, as their populations grow so fast. Many residents live in substandard housing, areas called shantytowns. Houses in shantytowns are built from scraps of wood, metal, and even cardboard, or whatever else the owner can find. They usually have one to three rooms, sometimes with a detached kitchen and an outhouse or pit latrine. Sometimes standpipes are available as a water source, but residents might have to buy their water from trucks that travel through these neighborhoods.

If the government does not have the shantytowns torn down because they are constructed on illegally possessed land, they may mature into more substantial housing with time. As members of the family find jobs they begin to make improvements, such as building rooms with cinder blocks and opening a small store in the house's front room. Researchers are now suggesting that shantytowns may help solve the housing problems of the largest Caribbean cities through their self-help nature. Some governments are helping these people provide for themselves by supplying electricity and water pipes to their neighborhoods. They also have built multistory public apartment units in cities like San Juan, Fort-de-France, Pointe-à-Pitre, and Charlotte Amalie (St. Thomas), but these have rarely provided solutions to the urban housing problem because many have become rundown and sometimes have become high-crime areas.

The largest Caribbean cities, those whose population exceeds 100,000, are becoming increasingly modern in appearance. Originally, even the British, French, and Dutch cities appeared similar to the typical Spanish colonial town. All the primate cities originated as colonial ports with gridiron street patterns focused around a central square or plaza. Usually, the wealthy lived as near as possible to the downtown central square, since it was here that greatest accessibility was afforded to public buildings, businesses, and face-to-face contact with important people. Today, however, automobile use is becoming more widespread and the largest cities especially are beginning to resemble cities in North America and Europe. The central business districts now have high-rise office and retail buildings. The areas surrounding the central business districts have often deteriorated, so they have become home to the poor instead of the rich. Meanwhile, the wealthy and middle classes have moved to the suburban periphery, just as they have in large U.S. cities, and the social gradient has reversed itself from what it was during colonial times. Also, in some of the largest metropolitan areas like San Juan and Kingston, shopping centers have developed in the suburbs.

FOLK HOUSING

The town was always more of a focus of activity for the Spanish and Dutch Caribbean territories than it was on the British and French islands. That is, most Spanish and Dutch colonialists preferred to live in the few urban settlements and not on dispersed farmsteads. The preference for living in town was based on the lateness of the arrival of the plantation system as the dominant economic sector to the Spanish colonies and to the mercantile orientation of the Dutch islands. As a consequence, the Spanish towns, in particular, were more grand and carefully laid out than their British and French counterparts. Spanish urban architecture in the Caribbean was generally more impressive, as reflected particularly in the churches they built. Dutch settlements differed somewhat from the Spanish because their emphasis on commerce made warehouses, port facilities, and the counting house (for taxes) more characteristic landmarks than churches. Often Dutch merchants lived in the rear or in an upper story of their business establishments, much as they did in Amsterdam. The Spanish merchants were likely to have more impressive homes than the Dutch, separated, but not far away, from their businesses.

The British and French colonialists, on the other hand, preferred to live on their large rural estates. The wealthiest owned the plantations and their stone mansions and mills persist as artifacts of the past on the rural landscape. While the Spanish, and to a lesser extent the Dutch, excelled in the cities, they had nothing to compare with the great houses in rural areas of colonies such as British Jamaica and French St. Domingue (Haiti).

The truth is, however, that most of the British and French did not own large estates and this is even more true of the former slaves who used to work on the plantations. The types of housing these poorer masses lived in evolved slowly since the first settlers began arriving in large numbers during the seventeenth century. From the post-emancipation period until the mid-1950s thatch-and-wattle houses were prevalent throughout the British and French islands. Building material for these houses consisted mainly of guinea grass, brittle thatch, tyre palm, birch berry tree limbs, and coconut palm fronds. After the house was framed with wooden poles, its roof was finished with dried guinea grass which was sewn together with large wooden needles. This finish would last for three to five years, unless a hurricane occurred, after which time the roof was replaced. The supple birch berry tree limbs were used to frame the house by weaving them together in a typical pattern, and then they were coated with a mixture of white lime and mud (a forerunner of today's cement), giving the houses a white stucco appearance. Usually, these homes had one or two rooms with a detached shed that served as a kitchen. In many respects (except for their stucco-like finish) these houses were similar to the homes built by the Arawak Indians before the arrival of the Spanish. The Arawaks called their houses *bohíos*. The Spanish rural settlers often adapted the bohío for their use, and some are still seen in the rural landscapes of the poorer territories like the Dominican Republic and Haiti, and even in remote areas of Cuba.

By the middle or late nineteenth century thatch-and-wattle housing was being replaced by wood frame houses sided with either boarding or shingles. The chattel houses still seen in Barbados and the Windward Islands are representative of this style. Often the families that lived in these structures owned the buildings but not the land they occupied. Sometimes the houses were moved if the property owner decided to use the land for another purpose.

Massive poured concrete homes and buildings also began to appear by the early twentieth century, especially after several severe hurricanes struck the islands. In the 1940s board-sided houses with galvanized or zinc roofing became common. But it should be noted that these were not distinct architectural phases because several housing styles coexisted at any one time. The type of house depended on the owner's financial and social status, as well as the availability of building materials. A typically modest board house consisted of one room with the sitting and sleeping areas separated by a room divider. Those with higher incomes had several rooms and often a large covered porch that became a social focal point for the family. The room dividers were constructed of pieces of board, used crates, driftwood, or even cardboard. These would be decorated with paper, which might include catalogue pages or other pieces of attractive paper, stuck on the divider with flour paste. Some of the more elaborate dividers contained lattice work which was ornamental and also allowed better air circulation during humid, warm, tropical nights.

The floors of the homes consisted of sand, dirt, or board depending on finances. Windows were mainly wooden with shutters that latched on the inside. In wealthier homes, lattice screens or jalousies were added to afford privacy and air circulation. The hip roof with a gutter running around it became popular because it was useful in collecting rainwater on the drier islands. Water was diverted from the roof via pipes and drains to barrels or a cistern. Hip roofs are still very common throughout the Leeward Islands and the Virgin Islands.

Foundations were not dug into the ground as is done today. Instead the house often sat upon wooden, brick, or concrete block pillars to make it more difficult for insects and rodents to enter the home. Also, houses elevated this way allow somewhat better air circulation. Houses built on pillars are still very common in all the Lesser Antilles. Entry to them is usually via steps, which might include a large stone, log, wooden crate, or crude wooden steps.

On the French islands it is especially common to see two-story wooden houses with the top floor overhanging the bottom one. The bedrooms are located on the upper floor where they can take advantage of the cool breezes during the night with their windows open. Ground-floor windows are almost always closed to ensure privacy. The overhang provides a rain-protected porch in front of the ground floor for socializing.

In poorer rural areas throughout the Caribbean, where there is no inside plumbing, kitchens still are often separated from the house, as are outhouses. Meals are frequently cooked on iron outdoor grills known locally in the British islands as coal pots. Often these modest homes are decoratively painted with several colors such as bright pinks, turquoises, and pastels both inside and out. In rural areas the houses are frequently surrounded with a small garden, called door gardens in the British Caribbean.

Since the 1950s more houses have been built with both poured concrete and locally manufactured concrete blocks. The hollow spaces in the blocks protect against the tropical heat, make them lighter to transport, and reduce the cost of construction compared with solid concrete walls. Also, today both electricity and inside plumbing are becoming commonplace throughout the Caribbean, especially on the wealthier islands. But throughout much of Haiti and the poorer and more remote areas of some other islands there are homes that still lack these amenities.

The observant visitor can recognize many of these different building types. Vernacular housing exhibits the same type of diversity that characterizes everything else in the Caribbean.

PART TWO

The Itinerary

Caribbean Cruise

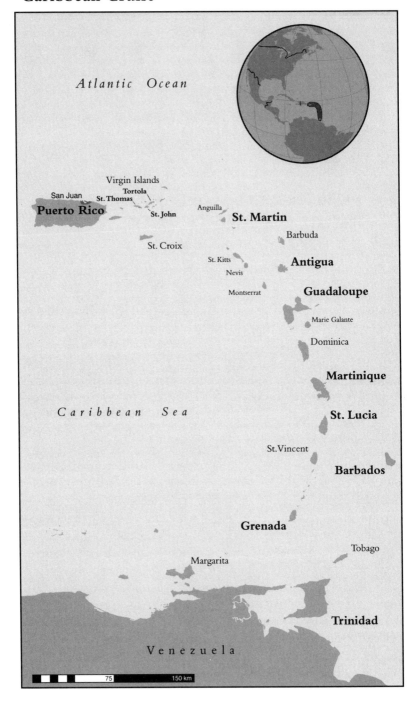

△ *Day One*

PUERTO RICO—LAND OF SPANISH INFLUENCE AND CHANGE

Among all the islands of the Caribbean none better exhibits Spanish influence than Puerto Rico, even though the island's economic destiny, since 1898, has been tied to its political affiliation with the United States. It was discovered by Columbus during his second voyage in 1493 and for the next four centuries it remained a Spanish colony. The language (Spanish) spoken by its people, its majority religion (Catholicism), its architecture, its holidays, and virtually every other aspect of its culture reflect this long period of Spanish domination. Since the late 1950s, the Spanish architectural integrity of the original center of San Juan has been restored through one of the world's most ambitious and vigorous historical preservation programs. A journey to the rural settlements of Puerto Rico's central mountains provides a visitor with a vivid impression of typical Spanish colonial towns, complete with an attractive central plaza, a nearby church dominating the town's architecture, a gridiron street pattern, Spanish stucco-styled houses, and cobblestoned and brick-lined streets. A hike into the surrounding countryside provides a glimpse of colonial-style tobacco and coffee farms and the home of the legendary *jibaro* (peasant farmer) who became the symbol of Puerto Rico's poor agrarian class.

In addition to its Spanish flavor, another outstanding characteristic of Puerto Rico is that few countries in the world have experienced as much change during the past half century as it has. Certainly none of the islands in the Caribbean, with the exception

Puerto Rico

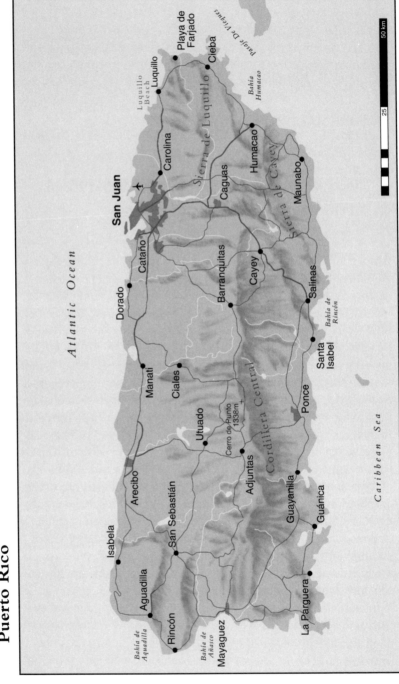

of Cuba, has witnessed such a human transformation. For instance, it was common as recently as the 1940s to refer to Puerto Rico as the poorhouse of the Caribbean. That is no longer the case, because the island has developed to the point that it now has the highest standard of living of any of the former Spanish-colonial possessions in Latin America. In the Caribbean only the Bahamas, Barbados, the U.S. Virgin Islands, the Cayman Islands, the Netherlands Antilles, and the French West Indies have standards of living as high or higher than that of Puerto Rico.

In 1940 the per capita gross national product of Puerto Rico was a paltry $154, but by 1989 it had risen spectacularly (by a factor of almost forty times) to $6,060. Even when standardized for inflation during this period, the real growth in per capita gross national product was by a factor of 5.4. During this period the economy had been transformed from one that was dominated by agriculture, especially the production of sugarcane, to one that is more diversified and reliant upon manufacturing, tourism, and services. Today ninety-seven percent of all homes on the island have television sets; one of every four households has a VCR; there is one car for every three Puerto Ricans and one telephone for every four persons. Life expectancy at birth has risen from forty-six years in 1940 to seventy-four years (compared to with seventy-five years for the United States) and illiteracy has been reduced to about ten percent.

The social and economic growth that has characterized Puerto Rico has taken place largely since 1948. In that year the Puerto Rican and U.S. governments embarked upon a development program known as Operation Bootstrap. Its primary goal was to promote industrialization through industrialization by invitation. Investment capital was attracted from U.S. firms by offering them tax incentives. In addition, the Puerto Rican government built buildings and industrial parks for branch plants of U.S. corporations and helped them train their Puerto Rican workers. For a while the lower labor costs of island workers was also a major factor attracting such labor-intensive industries as textiles and apparel. But as wages have risen on the island, emphasis has changed to high technology industries such as pharmaceuticals, electronics, petrochemicals, and scientific instruments.

Physical Features

Puerto Rico is the smallest and easternmost of the Greater Antilles, but it is still almost twice as large as Trinidad and Tobago, the next largest Caribbean country discussed in this guidebook. It is slightly over 100 miles (160 kilometers) long and about 35 miles (56 kilometers) wide, with a total area of 3,435 square miles (8,768 square kilometers). In addition to the main island, Puerto Rico includes the two eastern islands of Vieques and Culebra and the western island of Mona. It also includes a number of smaller islands and keys of lesser consequence.

About seventy-five percent of Puerto Rico is hilly or mountainous. With a population of 3.5 million persons it has a person/land density of approximately 1,000 people per square mile (386 people per square kilometer), but its density in terms of agriculturally usable land (called the physiological density) is much higher, exceeding 4,000 per square mile (1544 per square kilometer). Thus, despite the fact that some food crops, sugar, coffee, citrus crops, and a little (less than 1,000 acres or 1,620 square hectares) tobacco are still grown on the island, between one half and three quarters of its food requirements are imported in any given year.

The mountainous backbone of the island is comprised of the Cordillera Central in the west and center and the Sierra de Cayey and Sierra de Luquillo in the east. Average elevations are between 2,000 and 4,000 feet (610 and 1,220 meters), with the highest elevation being 4,389 feet (1,338 meters) at Cerro de Punta in the Cordillera Central. The remainder of the island consists of lowlands, such as the Caguas Valley, and a narrow ribbon of coastal plains that reaches a maximum width of about 13 miles (20 kilometers) on the island's northern side.

The relief of the island gives rise to sharp contrasts in precipitation on the windward and leeward sides of the island. The northeastern Trade Winds produce a much wetter environment on the northern side of the island. For instance, while the northern lowland city of San Juan receives an average of about 60 inches (150 centimeters) of annual rainfall, the figure on the northern slopes of the mountains is well above 100 inches (250 centimeters), and in the extreme northeastern corner of the Sierra de Luquillo, known as El Yunque, annual precipitation often exceeds 200 inches (500 centimeters) and a nearly rain-forest vegetation pattern prevails. However, the southern side of the island is not as wet. Here annual rainfall averages between 30 and 40 inches (75 to 100 centimeters), and the high tropical evaporation rate produces dry conditions with consequent grasses and arid vegetation types, such as several varieties of cacti.

Since pre-Columbian times the physical features of the island, like its human conditions, have also changed greatly. Because of agriculture and grazing activities, only about one percent of the land is still under virgin forest, although forests do exist in some mountainous areas. Rivers have been dammed into lakes to provide a more reliable supply of water and to produce a small amount of hydroelectricity. Urban expansion has also encroached upon the native vegetation patterns. Puerto Rico does not have a rich natural resource base. There are almost no precious metals, except for some copper which has yet to be mined in significant quantity. Therefore, the economic progress the island has made has been achieved not because of rich natural resources, but despite the lack of them.

History

Christopher Columbus was the first European to discover Puerto Rico on 19 November 1493. He found it by accident during his second voyage westward from Spain to the island of Hispaniola. There were perhaps 30,000 Taíno Indians (a branch of the Arawaks) living on the island, but they disappeared within a century due to the ravages of disease and Spanish cruelty. Columbus renamed the island from the Indian name of Boriquén to San Juan Bautista, after St. John the Baptist. For short, it became known simply as San Juan.

Fifteen years later, in 1508, Juan Ponce de León established the first Spanish settlement about 3 miles (5 kilometers) inland from San Juan in Caparra. In official records it became known as

Ciudad de Puerto Rico, or more simply Puerto Rico. Thus, ironically, the early names for the island and its main city were reversed from what they are today.

Because of mosquitoes and poor drainage, the original settlement at Caparra was moved in 1521 to the current site of Old San Juan on San Juan Island. The new town was designed according to the system specified by the Laws of the Indies, a scheme that was to be applied to most new Spanish settlements in the New World. A central plaza, the Plaza de Armas, served as the focus of activity and a grid pattern of streets intersecting at right angles was laid out. Surrounding the plaza were important government administration buildings and nearby was the cathedral.

Puerto Rico remained a Spanish colony until the Spanish–American War of 1898. With the defeat of the Spanish, the United States obtained the island (as well as Cuba and the Philippines). In 1917, under the Jones Act, Puerto Ricans became citizens of the United States. In 1948 the first Puerto Rican was elected to the island's governorship. In 1952 the island became officially known as the Commonwealth of Puerto Rico, freely associated with the United States.

The island's commonwealth status places it in a confusing position between statehood and independence with respect to the United States. For instance, Puerto Ricans cannot vote in U.S. federal elections as long as they reside on the island. However, if they move to the mainland they can vote in U.S. elections. While living in Puerto Rico they can vote in local elections and they do not pay federal U.S. income taxes, but they have complete freedom to move back and forth between the island and the United States because they are U.S. citizens. Taxes paid on rum and distilled spirits exported to the United States are returned to the Puerto Rican government to be spent on the island. The U.S. government provides for the island's defense and determines its foreign policy. It also subsidizes the Puerto Rican economy both directly through such transfer payments as food stamps, welfare expenditures, and other grants and indirectly through tax concessions. U.S. government transfer payments alone amount to almost $2,000 annually for every island resident.

Perhaps the most emotional political issue in Puerto Rico today is its relationship with the United States. Basically, the three choices appear to be continued commonwealth status, statehood, and independence. The island's three main political parties have platforms representing each of these positions. The pro-statehood and pro-commonwealth parties usually receive the largest number of votes in elections, swapping power from time to time. The independence party, though vociferous, usually receives less than ten percent of the votes.

Metropolitan San Juan, one day

In 1980 the U.S. Census Bureau defined the San Juan metropolitan area to include nine municipios. These units are similar to counties in the United States. Preliminary results from the 1990 census indicate that together these municipios contain about 1.2 million inhabitants, or approximately thirty-three percent of Puerto Rico's entire population. Among the cities on the twelve islands covered in this guidebook, San Juan is overwhelmingly the largest. In fact, among all the Caribbean islands, only Havana and Santo Domingo have larger populations.

Greater San Juan is made up of a number of districts contained within its nine municipios. Two are found on San Juan Island: San Juan Antiguo (Old San Juan) and Puerto de Tierra (Doorway to the Land). To the east of San Juan Island is a peninsula separated from the island's mainland by three bodies of water: San Juan Bay, Martin Peña Canal, and San José Lagoon. Among its most important districts are Santurce (a major retail business district), Condado (the main area of highrise hotels and exclusive condominiums), Miramar (an area of hotels and apartment complexes south of Condado Lagoon), and Isla Verde (the main beach of San Juan, which is occupied by hotels and is next to the Luis Muñoz Marín International Airport). The districts contained on the mainland south of the peninsula are usually considered to be the suburbs and exurbs of metropolitan San Juan.

Old San Juan

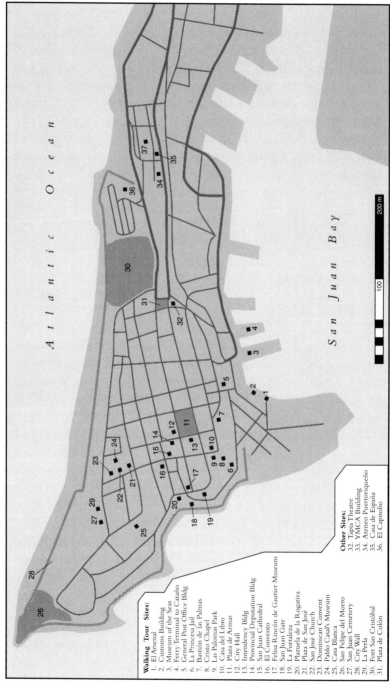

Walking Tour Sites:
1. El Arsenal
2. Customs Building
3. Museum of the Seas
4. Ferry Terminal to Cataño
5. General Post Office Bldg
6. La Princesa Jail
7. Bastión de las Palmas
8. Cristo Chapel
9. Las Palomas Park
10. Casa del Libro
11. Plaza de Armas
12. City Hall
13. Intendency Bldg
14. Provincial Deputation Bldg
15. San Juan Cathedral
16. El Convento
17. Felisa Rincón de Gautier Museum
18. San Juan Gate
19. La Fortaleza
20. Plazuela de la Rogativa
21. Plaza de San José
22. San José Church
23. Dominican Convent
24. Pablo Casal's Museum
25. Casa Blanca
26. San Felipe del Morro
27. San Juan Cemetery
28. City Wall
29. La Perla
30. Fort San Cristóbal
31. Plaza de Colón

Other Sites:
32. Tapia Theatre
33. YMCA Building
34. Ateneo Puertoriqueño
35. Casa de España
36. El Capitolio

Atlantic Ocean

San Juan Bay

200 m
100

The two field trips that have been designed for Puerto Rico fit the theme of change. They both involve the metropolitan area of San Juan because one of the most important social changes that has accompanied Puerto Rico's economic growth has been the urbanization of the Puerto Rican population, as people have moved from the countryside into the cities. In 1940 only about forty percent of all Puerto Ricans lived in urban places. Today the figure exceeds seventy percent. In fact, one in three Puerto Ricans live in the metropolitan area of San Juan alone. The first field trip involves a walking tour through the historically preserved section of Old San Juan and illustrates what urban life was like in Puerto Rico centuries ago. The second is a driving trip through the newer business districts and suburbs of metropolitan San Juan and illustrates the island's twentieth-century urban life-styles.

WALKING TOUR OF OLD SAN JUAN, THREE TO FOUR HOURS

Old San Juan (San Juan Antiguo in Spanish) comprises a seven-square-block area beginning at its eastern end with the Plaza de Colón and extending to San Juan Island's western end. This is the part of the city that originally was enclosed by a wall for protection against British, French, Dutch, and pirate invaders. In the 1950s it was designated a National Historic Zone. Since then, it has been reconstructed and renovated under the watchful eye of the Puerto Rican Institute of Culture to ensure that all changes blend with the Spanish colonial architectural style of this area.

Old San Juan is appropriately named because many of its buildings were constructed 200 to 300 years ago, some even dating to the sixteenth century. It became a charming residential and commercial district with narrow streets, buildings with balconies, and iron grillwork around the windows. Most of the homes here look plain from the outside. In typical Spanish colonial style, most are built right up to the sidewalk with no front lawns. But many have large interior patios graced with hanging plants and even fountains. Many of Old San Juan's streets are paved with bluish-gray bricks called adoquines, which, beginning in 1883, were made from slag derived from European iron foundries. Legend has it that these bricks were brought to the island as ballast in Spanish galleons, but historians have disproved this.

This trip has been designed as a walking tour because of the area's narrow streets and frequent heavy traffic. It covers a distance of approximately 3 miles (5 kilometers), but the visitor should reserve about three to four hours for the excursion because of the temptation to stop and enter many of the sights along the way. The trip begins at the cruise-ship docks along Calle Marina on Old San Juan's southern side, next to San Juan Bay. The first seven stops are within 300 yards (300 meters) of the docks and the visitor is advised to simply browse around this very interesting area. The structured itinery begins with site number eight. Notice that the Spanish words *calle* and *avenida* mean street and avenue in English.

The *Museum of the Seas,* on Pier 1 next to Calle Marina, contains model ships, maps, maritime instruments, photographs, and plates commemorating San Juan's history as a port. The ferry leaves the *Ferry Terminal to Cataño,* on Pier 2, for Cataño and Hato Rey. *Plaza de Hostos* is where Calle Marina intersects Calle San Justo. The plaza is named after a well-known nineteenth-century writer who espoused independence for Puerto Rico. Public cars wait to be rented here by tourists disembarking from the cruise ships. The drivers pass the time of day playing dominoes. On its northern side is the old *Banco Popular Building,* built in Art Deco style during the 1930s. The *Customs Building* is immediately to the west of Pier 1 on Calle Puntillo. *El Arsenal,* at Calle Puntillo south of the Customs Building, was a naval station built by the Spanish in 1800 and used by shallow-draft vessels to patrol the lagoons and channels in the mangrove swamps around San Juan. The Spanish made their last stand against the American forces here in 1898 during the Spanish–American War.

The *General Post Office Building* is on Calle Recinto Sur where it intersects San Justo and Tanca streets. *La Princesa Jail,* on Paseo de la Princesa at the bottom of the part of the old city wall beneath Cristo Chapel, was built in 1837 as a jail. It has now been restored and contains offices of Puerto Rico's Tourist Department.

Follow Calle Marina westward to Calle Tacna. Turn north on Tacna and proceed to Calle Recinto Sur and turn left. Follow

Calle Cristo, Old San Juan, Puerto Rico. Photograph by Thomas D. Boswell.

Recinto Sur for about 200 yards (200 meters) until it ends at Calle Tetuán. On the southwestern corner of this intersection is a small park, *Bastión de las Palmas.* Originally built for the emplacement of guns, this fortification was planted with trees by a civic group and now serves as a small park that offers a superb view of San Juan Bay.

Follow Calle Tetuán westward for about 100 yards (91.5 meters) until in ends at Calle Cristo. *Cristo Chapel* is at the southern end of Calle Cristo at this intersection. It was built in 1753 to commemorate a tragedy that occurred during the saint's day festivities for San Juan. Among the events held was a horserace down Calle Cristo. According to legend, one young rider failed to negotiate the turn onto Calle Tetuán and, plunging over the cliff, saved himself by holding onto a tree branch. Historians now believe that the rider in fact met his death. The small chapel currently blocks off the end of Calle Cristo so a similar tragedy will not happen again. Look inside the doorway of the chapel to see its small silver altar.

Las Palomas Park, next to the western side of Cristo Chapel, at the western end of Calle Tetuán. Hundreds of pigeons make this park their home. Most are tame enough to land on the hand of people who come here to feed them. On weekends children are bused to this park from all over the island. The park is perched on top of the southern extension of the old city wall and affords a beautiful view of San Juan Bay below. On a clear day the interior mountains of Puerto Rico can been seen from here.

Turn north on Calle Cristo as you leave Las Palomas Park. *Casa del Libro* is located on the eastern side of the street, only a few doors away from the Cristo Chapel at 255 Cristo. This is an eighteenth-century house used today as a museum and library devoted to the art of printing and book-making. It holds a fine collection of old pre-sixteenth-century masterpieces.

Travel north on Calle Cristo and turn right on Calle Fortaleza. Proceed one block east to Calle San José and turn north. The *Plaza de Armas* will be on your right side about one block north. This was San Juan's main square when it was laid out in the sixteenth century. It was used as a military drill ground when preparing for attacks and this is where it derived its name. It is now a social gathering place. The four statues that preside over the plaza represent the four seasons and are over a hundred years old.

City Hall is on the northern side of the Plaza de Armas. Called in Spanish the Alcaldía, it was constructed in stages between 1604 and 1789. Its double arcade, flanked by two towers, resembles Madrid's city hall. The formal assembly room, which occupies most of the second floor, was for many years the focus of San Juan's social life.

The *Intendency Building* is on the western side of the Plaza de Armas. This three-story building contains a large interior patio and an ornate neo-classical facade. It is considered to be one of the finest works of nineteenth-century Puerto Rican architecture. It was home of the Spanish Treasury from 1851 to 1898. During the twentieth century it has housed at different times Puerto Rico's Departments of Interior, Education, Treasury, and Justice. It currently is the home of the commonwealth's State Department.

The *Provincial Deputation Building,* at the northwestern corner of the Plaza de Armas on the corner of San José and San Francisco streets, housed Puerto Rico's first representative body, which came into existence in the latter half of the nineteenth century. Before the building's construction, this site was the location of San Juan Cathedral's cemetery. After that it served as a marketplace. In 1897 Puerto Rico gained autonomy from Spain and on 17 July 1898, the first insular parliament opened its offices in the Deputation Building. Eleven days later the U.S. marines invaded the island and autonomy ended as the U.S. army took over governance of the island. The building has been restored recently and now houses some of the offices of the State Department.

Walk one block west on Calle San Francisco and then turn north on Calle Cristo. The cathedral is about half a block north on Calle Cristo, on the eastern side of the small *Plazuela de las Monjas* (Nuns' Square). The first version of this church was built in the 1520s with wood siding and a thatched roof. In 1540, a hurricane destroyed it and it was reconstructed. During the early nineteenth century it was largely reconstructed a second time after the British troops of the Earl of Cumberland looted and badly damaged it. The last restoration of the cathedral was completed in 1977. Since 1913 the remains of Juan Ponce de León have lain here in a tomb near the transept of the building's north side.

El Convento, near the cathedral on the northern side of the Plazuela de las Monjas, was a convent established for Carmelite nuns. It was built over 300 years ago in 1651, but now it is a hotel run by Ramada Inn. *Casa Cabildo,* on the western side of the Plazuela de las Monjas, was built in 1521 and served as San Juan's original city hall. Now it houses an interior design company.

Felisa Rincón de Gautier Museum, on San Juan Street near where it intersects Calle Recinto Oeste, was once the home of Felisa Rincón de Gautier, who was the dynamic and controversial mayor of San Juan from 1946 to 1968. This is a typical eighteenth-century-style Spanish home.

San Juan Gate, on Calle Recinto Oeste, just west of the Gautier Museum, was part of the gate system that protected the city from sea-borne invaders. Its huge wooden doors opened on the little cove just north of La Fortaleza, where small sailing ships anchored in the early years of colonial rule. When high-ranking officials arrived, they were met here by local dignitaries and escorted up Calle San Juan to the cathedral, where a mass was offered in thanks for a safe voyage. When bigger ships came into use in the eighteenth century, such ceremonies were held at the San Justo Gate opposite the southern end of Calle San Justo near the cruise-ship piers.

La Fortaleza, south just off Calle Recinto Oeste overlooking San Juan Bay, was authorized by Charles I as a defense against Carib Indians raids. It was completed in 1540. But the fortress soon proved to be of little military value and it became the official residence of the colony's governors. In fact, it is the home of the governor of Puerto Rico. This is the oldest executive mansion still used as such in the Western Hemisphere.

At the western end of Calle Monjas stands a statue of a priest followed by three women, donated to the city of San Juan in 1971 to commemorate its 450th anniversary. Placed in a small plaza, *Plazuela de la Rogativa,* beside the city wall, it refers to the story of a miracle that occurred in 1797. British troops were besieging

San Juan and saw unusually heavy activity one night. They believed reinforcements had arrived in San Juan to assist in a Spanish defense. As a consequence the British forces withdrew. Actually what they saw, but did not understand, was a *rogativa,* a religious procession of women carrying torches and singing as they followed their bishop.

Travel north on Calle Recinto Oeste to Calle Hospital, turn east, walk for about half a block, and look to your left. This is the way many streets used to look in San Juan before becoming modified to suit vehicular traffic. Because the streets were for pedestrians, they often used step streets. These stairs were probably built during the late sixteenth or early seventeenth centuries.

Continue walking eastward on Calle Hospital and then turn north on Calle Cristo. Walk north on Cristo until it intersects San Sebastian. Turn right on Calle San Sebastian and look to your left. The small square is the plaza you are looking for. *Plaza de San José* has become a favorite meeting place of both the young and old. The statue in the square is of Ponce de León and was made from British cannons captured and melted down after Sir Ralph Abercromby's unsuccessful attack on San Juan in 1797.

San José Church is on the north side of the Plaza de San José. Construction was begun in 1532 on this, Puerto Rico's oldest church and the second oldest in the Western Hemisphere. (At least this is true if you consider the cathedral of San Juan to have begun construction in 1540, not during the 1520s when its first wooden version was constructed.) This was the church for Ponce de León's descendants for several hundred years. When Ponce was killed by Seminole Indians in Florida in 1521, his body was buried here until 1913, when it was moved to San Juan Cathedral.

The *Dominican Convent,* north of the plaza and immediately behind San José Church, is a large structure that currently serves as the offices for the Institute of Puerto Rican Culture, the agency that oversees the restoration of Old San Juan. Its construction was begun in 1523. While originally serving as a convent, it later became popular with foreign armies. For instance, the Earl of Cumberland used it during the brief British occupation of San Juan in 1598. The Dutch invader Bowdoin Hendrickson quartered some of his troops here during the Dutch occupation. Finally, it was used as the home offices of the U.S. Army's Antilles Command from 1898 until 1966.

Pablo Casals Museum is on the eastern side of San José Plaza. This world-renowned cellist moved from Spain to Puerto Rico in 1957 and lived here until his death in 1973. The museum contains various memorabilia, such as videotapes of performances, manuscripts, musical instruments, and photographs.

Walk westward along Calle San Sebastian to its end (less than two blocks). The white house behind the gate is *Casa Blanca.* It was to be built as a gift for Ponce de León, but he was killed before it was completed. His descendants lived in it for about 250 years until it was taken over by the Spanish military as the residence of its commanders. Between 1898 and 1966 it served as the residence of the U.S. miliary commander in Puerto Rico. Following restoration, it has since served as a museum of sixteenth- and seventeenth-century family life in Puerto Rico.

Follow the road that runs in front of the Casa Blanca, northwestward to the fort, *San Felipe del Morro.* It is usually called El Morro for short. This is probably the most photographed building in the Caribbean. Its construction began in 1540 when the fortifications at La Fortaleza proved inadequate. It is comprised of six levels and stands 140 feet (43 meters) above San Juan Bay, which it was designed to guard. A tour of the facilities is a must for any serious visitor to San Juan. It has been designated both a World Heritage Site and a National Historic Site, which is administered by the U.S. National Park Service. Orientation and slide programs are available in both Spanish and English.

San Juan Cemetery, adjacent to the north wall of El Morro, between the fort and the Atlantic Ocean, was inaugurated in 1814 and serves as the final resting place for some of Puerto Rico's most prominent citizens. It contains many elaborate tombstones and a circular chapel built and dedicated to St. Mary Magdalene in 1863.

Construction of *the City Wall* surrounding Old San Juan, adjacent to *Norzagaray Street* on the seaward side, began in the 1630s. It is composed of two forty-foot-high (twelve-meter-high) sandstone parallel walls, whose stones are held together by cement.

La Perla (shantytown), San Juan, Puerto Rico. Photograph by Thomas D. Boswell.

The space between the two walls was filled with sand and then covered by bricks which slope downward toward the sea. The wall is approximately twelve feet (three and a half meters) wide at the top and twenty feet (six meters) wide at its bottom. At sundown, access to the city was cut off by closing its six heavy wooden doors.

La Perla is San Juan's oldest and most famous slum. It adjoins the eastern edge of San Juan Cemetery. Because of its view of El Morro and the Atlantic Ocean it is sometimes referred to as the prettiest slum in the world. La Perla originated as a squatters' settlement or shantytown during the late nineteenth century. Gradually its homes have been improved, but it is still an area characterized by poverty, unemployment, and a high crime rate. This is the slum Oscar Lewis wrote about in his classic work *La Vida*.

Fort San Cristóbal, at the eastern end of Calle Norzagaray and parallel to Muñoz Rivera Avenue, was built in 1771 as a partner in defense of Old San Juan with El Morro. Its purpose was to guard the landward side of this city. It is comprised of one main structure and five other independent units. It was considered a masterpiece in military design because the enemy could only enter the main fortress after capturing the other five individual structures, which were all connected by tunnels. Its museum is open to the public.

Plaza de Colón, at the eastern end of Old San Juan, was renamed in honor of Christopher Columbus in 1893, commemorating the 400th anniversary of his discovery of Puerto Rico. Originally it had been dedicated to St. James. In it is a statue of Columbus. The plaza now serves as a terminal for buses plying between Old San Juan and the Puerto Rican mainland.

Follow *Calle San Francisco* westward from Plaza de Colón to Calle San Justo and then turn south on the latter to return to the cruise-ship terminals where this trip started. San Francisco is one of the main shopping streets in Old San Juan. It is full of retail establishments, restaurants, and night spots.

New San Juan and Suburbs, three to four hours

New San Juan, as the term is being used here, refers to all of the metropolitan area of San Juan that is not included in Old San Juan. Some of New San Juan is not really so new. For instance, the town of Río Piedras was founded in 1714. Still, most of New San Juan has been built since 1900, so relative to Old San Juan it is new. In many ways it is a different world from Old San Juan because it is mainly characterized by a suburban landscape. Single-family homes prevail, with occasional areas of apartments and condominiums. Here lawns are the norm, unlike in Old San Juan.

The field trip to New San Juan covers about 25 miles (40 kilometers), so this excursion must be driven. Most people will

Suburban San Juan Tour

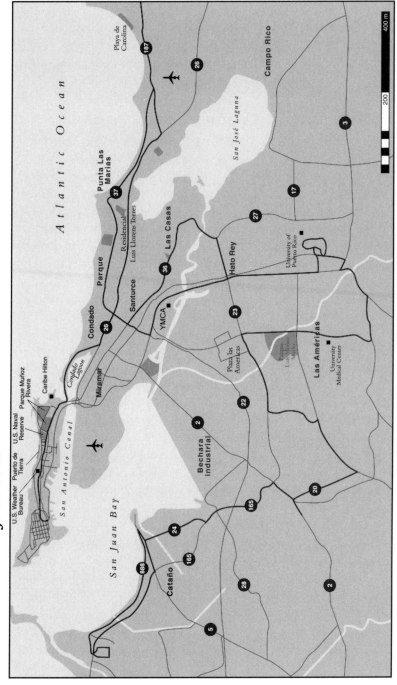

find that it takes between three and four hours to complete, depending on traffic, the day of the week, and the number of times the traveler stops. Like the old city trip, this one also starts at the cruise-ship docks of Old San Juan.

PLAZA DE COLÓN TO PUERTO DE TIERRA, TEN MINUTES

Drive north on Calle Tanca and turn east on Calle Fortaleza. Fortaleza is the other major east–west-running shopping street in Old San Juan, in addition to Calle San Francisco. Proceed east on Fortaleza to the Plaza de Colón. To the left of the plaza you will see a yellow and white building, the *Tapia Theater.* It was financed by subscriptions and by taxes on bread and imported liquor. Construction began in 1832 and continued off and on until the last quarter of the nineteenth century. One of the oldest currently used theaters in the Western Hemisphere, it was named after Alejandro Tapia y Rivera, who was one of Puerto Rico's most prominent playwrights. The theater was restored to its original architectural lines by the City of San Juan in 1976. Since then it has become the main setting for theater, dance, and other cultural events in Old San Juan.

Once you leave the Plaza de Colón, traveling east, you enter the *Puerto de Tierra* section of San Juan Island. On this trip you should proceed eastward on Avenida Ponce de León. On the northern corner where Avenida Ponce de León enters the Plaza de Colón is located San Juan's first *casino,* built in the 1917 for the Casino de Puerto Rico, which was a high-society club. It is no longer used as a casino and has been renamed the Manuel Pavia Fernandez Government Reception Center. The Puerto Rican State Department now uses it for receptions.

The *YMCA building,* on the northern side of Ponce de León a couple hundred yards (meters) east of the casino, was built during the early part of this century and restored in 1991.

Ateneo Puertoriqueño, adjacent to the eastern side of the YMCA building, is an exclusive social club whose primary purpose is to promote the arts and humanities in Puerto Rico. *Casa de España* is a beautiful four-towered and blue-tiled building just east of the Ateneo Puertoriqueño. Once a high-society gentlemen's social club, it now is used for a variety of cultural events and social receptions.

El Capitolio, east of the Casa de España between Ponce de León and Muñoz Rivera avenues, was Puerto Rico's capitol building. It was built in 1925 and greatly resembles the U.S. Capitol Building in Washington, D.C. It also resembles the capitol building in Havana. On either side of Capitolio can be seen the two low rectangular buildings that house the offices of the Puerto Rican Senate and House of Representatives.

On the northern side of Ponce de León, east of Capitolio, can be seen the buildings of the old *School of Tropical Medicine* of the University of Puerto Rico (now moved to the new medical campus in Río Piedras). It currently serves as the offices of Puerto Rico's Department of Natural Resources. East of the former medical buildings are buildings used by the U.S. military. Included is the old edifice that used to house the *U.S. Weather Bureau,* and farther to the east are the grounds for a small U.S. Naval Reserve station.

Muñoz Rivera Park was originally designed as a small square park over half a century ago. It was later enlarged to its current spacious size with wide walks and resting areas. It was named after a famous Puerto Rican journalist, poet, and statesman. Periodic craft and cultural fairs are held here. Grafted onto its northern edge is the Sixto Escobar Park where many junior sporting events are held.

San Gerónimo Fort, on a small point of land just before you cross the bridge connecting San Juan Island with the *Condado* area to the east, was built during the late 1700s to prevent landward attacks on the island. It has been restored by the Institute of Puerto Rican Culture and houses a military museum, old maps, and other historical materials. The *Caribe Hilton* (late 1940s), adjacent to San Gerónimo Fort, is a historic site because it was the first hotel built by the Puerto Rican government in an attempt to promote the island's tourist industry. A new twenty-story tower was added to it in 1972 and the entire complex was renovated; now it is one of San Juan's most modern and luxurious resorts. Next to the Caribe Hilton is the Radisson Normandie, built in Art Deco style in 1939 to resemble the famous French oceanliner *Normandie.* It was closed for several decades and then was renovated and reopened in 1989.

CONDADO, TWENTY MINUTES

Veer to the left and take the northernmost bridge from Puerto de Tierra to the Condado area and follow Avenida Dr. Ashford eastward. You will pass a small public beach on the left and then enter the Condado's *concrete canyon* of high-rise hotels and condominiums. Here you will see a number of San Juan's most famous hotels including the Regency Hotel, the Condado Plaza Hotel and Casino, the Condado Beach Hotel, the San Juan Convention Center, and the Hotel La Concha. If you look farther eastward up Avenida Ashford you will see on the left the Dupont Plaza Hotel. This hotel is the one that had a fire in 1989 that spread to its casino, killing approximately 100 people. Sabotage was suspected because the incident took place during an emotional labor dispute.

After traveling approximately 1 mile (1.6 kilometers), veer right at a fork in the road onto Calle Luisa and continue until you see the "on ramp" for the Boldorioty de Castro Expressway (Route 20). Continue eastward toward the Luis Muñoz Marin Airport. About 1 mile (1.6 kilometers) after you have turned onto the expressway, you will see on your left *Residencial Luis Llorens Torres*. This public housing project, built during the 1950s, was one of Puerto Rico's first such undertakings. Government projects like this are called *residenciales* in Puerto Rico. Many consider the Llorens Torres area a white elephant because of its unkempt slumlike appearance and high crime rate.

Turn northeast (to your left) off Route 20 onto Avenida Gobernadores. This will take you onto Avenida Boca de Congrejos, north of the international airport. Continue driving east for about 1 mile (1.6 kilometers). This takes you along San Juan's most popular public beach, called Balneario de Isla Verde, but recently renamed *Playa de Carolina* because it has been taken over by the *municipio* (municipality) of Carolina. Public beaches in Puerto Rico are called *balnearios*. This area along the beach and near the airport is called *Isla Verde* (Green Island, although it is not an island). Turn around and drive back in a westerly direction and veer to the right onto Avenida Isla Verde (Route 37). Here you will see the hotels of Isla Verde.

Continue driving west for an additional 2.5 miles (4 kilometers) and along the way you will pass the upper middle-class residential areas of Punta Las Marias, Park Boulevard, and Ocean Park. Follow the main road and you will notice that its name has changed to Calle Loiza. Turn south (left) onto Calle José de Diego. After you pass under Baldorioty de Castro Expressway, you will see the relatively new Government Center complex to your right. The Puerto Rico Planning Board is in one of the two tall towers here. The planning board was started by the well-known geographer Rafael Picó during the 1940s. Continue southward for about 0.5 mile (0.8 kilometer) and drive up the ramp leading to Luis Muñoz Rivera Expressway (Route 1) and head east across Martin Peña Canal and into Hato Ray.

MARTIN PEÑA, TEN MINUTES

Turn left on Avenida F. D. Roosevelt (Route 23). You are now entering Puerto Rico's poorest area, called *Martin Peña,* after the canal that runs through its heart. It is north (left) of the street you are on, next to Hato Ray to the south (right). Circle this area, but do not stop because this is a high crime section. Turn left on Calle Dr. Barbosa (Route 27—this is not the same Barbosa Expressway you were on earlier). Follow Barbosa for about 0.5 mile (0.8 kilometer) to Avenida Borinquen and turn left. Follow this street westward until it merges with Avenida Ponce de León about 1 mile (1.6 kilometers) later. The Martin Peña slum used to be much worse than it is now. The whole area in the 1960s and early 1970s was comprised of shacks forming a shantytown in this area. During the rainy season the area usually flooded; many of its houses were built on wooden stilts. Raw sewage was deposited in the canal and its odor became famous. During the 1970s and 1980s the Puerto Rican government cleaned up some of the more unsanitary sections of this neighborhood, but it is still a poverty-ridden area.

SANTURCE AND HATO REY, TEN MINUTES

Continue westward on Avenida Ponce de León. This takes you through the heart of *Santurce,* the main retail business center of

San Juan. Although this area was founded about a century ago as a fashionable suburb for Old San Juan, since the 1950s it has been converted to a business district along its two major east–west-running thoroughfares, Ponce de León and Fernando Juncos avenues. Now its residential communities have been converted to lower middle-class neighborhoods, as the housing has aged and filtered down to less affluent families. Even the business buildings are showing their age and much of the retailing activity that once characterized this area during 1960 to 1980 has fled to more distant suburbs and shopping centers.

Turn left on Calle José de Diego and retrace your path back to *Hato Ray* by turning east on Luis Muñoz Rivera Expressway (Route 1). Hato Ray is the undisputed high finance and banking capital of the Caribbean. In addition to Puerto Rican banks like the Banco Popular, there are many foreign banks here to take advantage of the communications system of San Juan and the financial security of an American territory. The area of most intense development along Route 1 is often called the *Golden Mile*.

RÍO PIEDRAS AND THE UNIVERSITY OF PUERTO RICO, TWENTY MINUTES

Continue southward on Route 1 to *Río Piedras*. Turn east (left) on Avenida Universidad (be careful because it is easy to miss). This will take you to the entrance of the *University of Puerto Rico*. This is the largest university in the Caribbean with about 45,000 students, of whom about 25,000 are enrolled on this campus. Drive around the beautiful grounds. The university's large octagonal clock-tower (built in 1937) has become a landmark in Río Piedras, as a reference point for giving directions.

After your visit on the UPR campus come back out the entrance to the university and turn south (left) on Avenida Ponce de León. Continue for about one third of a mile (a half kilometer) until you reach the beautiful central plaza of Río Piedras. Drive around it, or better yet get out of your car and walk around it. There is an underground parking garage beneath the plaza (an interesting contrast between the modern and traditional aspects of this plaza).

This square is distinguished by its Catholic church, gardens, benches, and older men playing dominoes. Public cars (called *públicos*) leave from here to go to all the towns on the island as an inexpensive form of transportation. A street off the plaza leads to the Plaza del Mercado, largest public marketplace on the island, whose wares include fruits, vegetables, and medicinal herbs. Río Piedras was founded as a separate municipio in 1714 but it joined with San Juan Municipio in 1951, so it is no longer a separate legal entity.

Return to Avenida Ponce de León and go south until it joins a freeway interchange. Follow the signs to Route 1 and then, almost immediately, turn off the freeway to enter (follow the sign) the University of Puerto Rico's Botanical Garden. This facility is on one of the university's experimental stations. It contains more than 200 species of tropical and subtropical vegetation and contains several beautiful ponds. Admission is free.

Get back on Route 1 heading west and drive about 0.5 mile (0.8 kilometer) until you can turn west (right) on Route 21. Continue for about another 0.5 mile (0.8 kilometer) and turn north on The Americas Expressway (Route 18). On the west side of the expressway you will see the facilities of the School of Medicine of the University of Puerto Rico. This is where University of Puerto Rico's School of Tropical Medicine moved. About a mile (1.6 kilometers) north, at the intersection with Route 23, on the left is a sports complex that contains the Hiram Bithorn Stadium and the Roberto Clemente Colosseum. Most of Puerto Rico's major professional sporting events are held here.

URBANIZACIONES, FIFTEEN MINUTES

Turn west onto Avenida F. D. Roosevelt (Route 23). Immediately on the right is the Plaza Las Americas, the largest shopping mall in the Caribbean, replete with fountains and flowered walks. It has many American stores—Sears, Toys-R-Us, J. C. Penney, and an interior restaurant emporium with booths selling foods typical of various countries around the world.

Follow Route 23 for about 1.5 miles (2.4 kilometers) and turn south on Avenida de Diego (this is not the same as Avenida José de

OTHER FIELD TRIPS

There are, of course, many other interesting field trips in Puerto Rico. If the visitor has more time and energy, we particularly recommend a trip to the beautiful beach of *Luquillo* and the fascinating rain forest of *El Yunque*. They are near each other on the island's northeastern end. Also highly recommended are driving trips to the western end of the island through the towns of Manatí, Arecibo, Aguadilla, and Mayagüez and to the island's second largest city, Ponce (with a population of almost 200,000). Another interesting trip is to the *limestone region* on the northern side of the island between Manatí and Arecibo. The best starting point for such a trip to see the karst topography of this region is Río Camuy Cave Park near the town of Lares on Route 129. Another beautiful trip involves driving via Route 10 through the heart of the Cordillera Central mountain range and passing through such picturesque colonial towns as *Utuado* and *Adjuntas*.

Diego in Santurce). This street passes through a middle-class *urbanización* that probably was built during the 1950s. Tracts of planned single-housing units (as opposed to apartments) are called *urbanizaciones* in Puerto Rico. Notice the commercial strip along this street. Turn right on Calle Delta and then left on Calle Escorial. These latter two streets will take you through a middle-class neighborhood typical of this area.

CAPARRA, FIFTEEN MINUTES

Turn right on Avenida Jesus T. Piñero (Route 17). Continue about a mile (1.6 kilometers) and turn right again onto Route 20 (it becomes R. Martínez Nadal Expressway). Continue another mile

(1.6 kilometers) and turn left onto Avenida J. F. Kennedy (Route 2). About 0.25 mile (0.4 kilometer) south on the right side are the remains of Ponce de León's original settlement at *Caparra*. It was from this site that Ponce moved in 1521 to start Old San Juan. There is a museum here and some of the foundations of the original buildings can be seen.

CATAÑO AND RETURN, ONE HOUR AND A HALF

Turn back north on Avenida J. F. Kennedy (Route 2) and drive about 0.5 mile (0.8 kilometer) and turn left on Route 165. After driving about 1.5 miles (2.4 kilometers) turn right onto Route 24 (Juan Avenida Ponce de León—not the same Avenida Ponce de León). This will take you into the municipality of *Cataño*, a relatively poor or lower middle-class community. Most who live here work elsewhere as blue-collar laborers. Many commute to work in Old San Juan on the ferry that leaves from the town's dock.

Follow Route 24 to its northern end and then turn west on Avenida Nereidas. Continue on this road westward for about 2 miles (3 kilometers) until you reach the grounds of the *Bacardi Rum Distillery*. Go in for a free tour of the world's largest rum plant.

Return to Cataño on Avenida Nereidas. If you are riding in a cab or bus, take the ferry back to Old San Juan. It only takes about 10 minutes for the ride and you arrive at the middle pier (Pier 2) next to the cruise-ship docks. If you are driving a rented car follow your map via the shortest route to where you are to return it.

△ Day Two

THE VIRGIN ISLANDS—
ISLANDS OF CONTRAST

The 130 or so islands, rocks, and cays (keys) that comprise the Virgin Islands are approximately 40 miles (64 kilometers) east of Puerto Rico. Geologically most of these islands represent a continuation of Puerto Rico, having evolved from volcanic processes longer ago than 25 million years. In fact, the two easternmost islands of Puerto Rico—Culebra and Vieques—are geologically part of the Virgin Islands, although administratively they are affiliated with the Commonwealth of Puerto Rico.

The volcanic activity of the Virgin Islands has long since ceased. It lasted just long enough to build a submerged mountain system— the emergent tops of which are the mountainous islands—that continues eastward from the Greater Antilles to the Anegada Passage. The elevations of the mountain backbones of these islands average about 1,000 feet (305 meters) in elevation, with the highest peaks being less than 2,000 feet (610 meters) above sea level. Only a few of the smaller islands (locally called islets), plus Anegada and St. Croix, do not fit this pattern because they are low-lying limestone and coral formations.

Administratively, the Virgin Islands are comprised of two parts, the U.S. islands (the USVIs) and the British islands (the BVIs). The USVIs have by far the larger population (about 105,000 compared with approximately 15,000 for the BVIs), and their economic development has progressed further. Conversely, the BVIs have favored a more cautious approach to their development

Abandoned sugar mill, St. Croix, U.S. Virgin Islands. Photograph by Thomas D. Boswell.

of tourism. For instance, the BVIs have no high-rise hotels or development strips as found in Puerto Rico and, to a lesser extent, in the USVIs.

Both groups of the Virgin Islands are affiliated with their respective home countries through semiautonomous arrangements that allow for most decisions to be made locally, similar to the situation in Puerto Rico. For instance, citizens of the USVIs are also U.S. citizens, but they do not vote in presidential elections, nor do they pay U.S. income taxes. The customs taxes paid on most items exported to the U.S. mainland are returned to the local government.

Both the USVIs and the BVIs depend heavily upon tourism as a generator of income. In both cases, the tourist sector provides more than half of personal income. Government transfer payments and employment are also important. The USVIs have exerted effort to attract foreign-sponsored manufacturing, especially on the island of St. Croix. But there is virtually no manufacturing in the BVIs, which are limited to the production of a very small amount of rum and a few food products. In addition to tourism, another growth industry in the BVIs is the new international business company industry. In 1984, legislation was passed to encourage international business companies to register their home offices here, the attractions being low taxes, use of the U.S. dollar as local currency (even though this is a British Crown Colony), and ease of registration procedures. The emphasis has been on the non-banking business because offshore banking and insurance facilities are well-provided for on other Caribbean islands, such as the Bahamas, the Cayman Islands, the Turks and Caycos Islands, and Barbados.

Agriculture, particularly the production of sugarcane, used to be the economic mainstay of the Virgin Islands. Almost all the native vegetation was cut over and for several centuries sugar remained king. But for a variety of reasons, including the emancipation of black slaves, soil depletion, steep slopes, and foreign competition, the sugar industry collapsed on most of the islands during the latter half of the nineteenth century. Although there is some production of food crops, such as vegetables and fruits, now most of the food consumed in the VIs must be imported, mainly from Puerto Rico, the Dominican Republic, and the U.S. mainland. Virtually everything else also must be imported, so prices here are at least fifty percent higher than in the United States for everything except liquor and jewelry.

Even water is at a premium in all the VIs because only about 40 to 50 inches (100 to 125 centimeters) of rain falls annually (except for a few higher elevations in the mountains). The steep slopes and thin soils cause most of the precipitation to run off into the surrounding sea. Almost all buildings are constructed with roofs designed to catch rainwater and channel it to underground cisterns. However, these efforts are not nearly enough to satisfy the need for water as the population of the islands and the number of visitors continue to grow. As a consequence, desalination plants have been constructed on St. Thomas, St. Croix, St. John, and Tortola, and the importation of bottled drinking water is common throughout the islands.

Both the U.S. and British Virgin Islands have living standards among the highest in the Caribbean, being higher than Puerto Rico, but comparable to those of the Bahamas, the Cayman Islands, and the Netherlands Antilles. Although some people are poorer than others, abject poverty is not widespread here as it is on some of the other Caribbean islands. Unemployment is very low, even lower than on the U.S. mainland and in Great Britain. In the USVIs some of the excess labor force that developed during the latter half of the 1980s has emigrated to the United States, most being comprised of people who immigrated to the USVIs at an earlier time. In fact, the USVIs is one of the very few territories in the world where the majority of the population is comprised of outsiders. Recent census figures indicate that only forty-five percent of its inhabitants were born in the USVIs. Almost thirty percent were born in other islands of the West Indies, while about thirteen percent came from the United States, and another five percent were from Puerto Rico (especially the island of Vieques). The remaining seven or eight percent came primarily from Europe or Latin American countries (other than Puerto Rico).

Although throughout most of their post-Amerindian history the USVIs were owned by Denmark, English has always been the

The Baths, Virgin Gorda, British Virgin Islands. Photograph by Thomas D. Boswell.

primary language. The reason is that even though the Danes governed the U.S. islands, most of the sugar plantations they established were run by English, Scottish, and Irish overseers. On the other hand the BVIs have been governed by the British since 1666. Although English is spoken here, the Virgin Islands have their own patois, called Calypso English or VI English Creole, which an outsider will often have difficulty understanding.

Almost all the population of the Virgin Islands lives on seven islands. The three main concentrations in the USVIs are St. Croix (50,000), St. Thomas (50,000), and St. John (4,000). In the BVIs the greatest population is on Tortola (about 12,000), Virgin Gorda (2,000), Anegada (500), and Jost Van Dyke (fewer than 500). Perhaps another thousand people live scattered throughout the rest of the islands. The field trips in this guidebook will involve Tortola, St. Thomas, and St. John.

Tortola: Go Slow Tourism, three hours

Tortola is the dominant island in the BVIs in every respect. In area (about 21 square miles, or 55 square kilometers) it is the largest of the BVIs, measuring approximately 11 miles (18 kilometers) in length and about 2 miles (3 kilometers) in width. It contains the highest mountain (Sage Mountain, 1,780 feet, or 543 meters), the largest population, and the capital of the BVIs, Road Town. It is the economic focus and tourism center for the BVIs. Despite these superlatives, it still maintains some aspects of a laid-back rural enclave, whose beauty and tranquility are its most notable attributes. But things are beginning to change in the BVIs, so much so that some Tortolians fear that the simple and peaceful life will be a thing of the past. As one well-known and long-time resident of the island said, "It is as if you can feel the island beginning to shake before a volcanic eruption."

Tortola has gone through at least four phases of land-use change. Shortly after the first Europeans (the Dutch) arrived they began cutting over most of the natural brush and forest cover to plant sugarcane. For approximately 150 years most of the island was covered with cane. But the emancipation of the slaves in 1834 brought about the demise of most sugar plantations within the next fifty years. This ushered in the second phase, which was a period when small farmers took over much of the land formerly owned by the large estates. They grew mainly subsistence crops such as vegetables, bananas, sweet potatoes, manioc (cassava), and various fruits. They also raised some livestock where the soil was less productive. But farming started to become less profitable during the early 1900s, as many residents of the BVIs traveled to Panama to help build the trans-isthmus canal and establish banana plantations along that country's Caribbean coast. Later, many Tortolian men and women were recruited in the Dominican Republic to cut sugarcane and be domestic servants. By the 1940s land-use patterns had changed a third time, as much (but not all) of the area devoted to crops was converted to pasture for the grazing of cattle. As workers were recruited during the 1960s and early 1970s to work in the USVIs, wages began to rise significantly in the BVIs.

Tortola, British Virgin Islands

Atlantic Ocean

Beef Island

Quaker Burial Grounds

Josiahs Bay

Paraquita Bay

Ridge Road

Fort George
Fort Shirley

Chapel Hill

Port Purcell

Road Bay

Kingstown

Prospect Reef Resort

Road Town

Albion

Sea Cow Bay

Nanny Cay

Brewer's Bay

Cane Garden Bay

Cane Garden Bay

Mt Sage 543m

Ft Recovery

Soper's Hole

Sir Francis Drake Channel

5 km

2.5

The grazing of cattle became unprofitable because of the increasing cost of hiring laborers to cut the rough older grasses, so they could be replaced by younger, more nutritious grasses that were better for the cattle. By the late 1970s and early 1980s the fourth phase of land use had begun. Now most of the land is unused, except for grazing a limited number of cattle, goats, and sheep. Both goats and sheep are more tolerant of poorer pasture than cattle, so they are more widespread throughout the island, whereas cattle tend to be restricted to the better grassland. A few farms are still scattered about the island, but now more than three quarters of the food consumed on Tortola is imported, as it is in most of the rest of the Virgin Islands. Nevertheless, the scars produced by the earlier land-use stages are clearly visible throughout most of the Tortola. Look for these on this island excursion.

PORT PURCELL TO BEEF ISLAND, THIRTY MINUTES

Begin at *Port Purcell,* where smaller cruise ships dock. Proceed eastward along *Waterfront Drive* (which after leaving Road Harbour toward the east becomes known as Blackburne Highway) through a number of small villages to the eastern end of the island, where Tortola is joined by a bridge to Beef Island. Along the way you will pass *Fort George* on the left. It was here that a plantation owner, *Arthur Hodge,* was hanged in 1807 for killing one of his slaves because he suspected the slave of eating a mango. This was a significant event because the cruelty Hodge bestowed on his slaves is often cited as one of the factors that ultimately turned many plantation owners on Tortola against the institution of slavery. The British emancipated slaves in the colonies twenty-seven years later.

Just beyond Fort George is *Baughers Bay,* where the ferry to Peter's Island and Norman Island dock. *Kingstown* is the site of the old capital of Tortola before it was moved to Road Town during the nineteenth century. Next, you will pass *Paraquita Bay,* which is a natural deep-water (300 to 500 feet, or 91 to 152 meters) embayment where yachts and other boats often take refuge during rough seas. This area is the largest amount of flat land on Tortola. Located here is the BVI Agricultural Station, where experimentation is conducted with new crops and agricultural techniques. Plans call for part of this land to be used for construction of the BVI's first college. Farther east along the road is *Hodges Creek,* which maintains Tortola's third largest marina.

About 0.5 mile (0.8 kilometer) north and to the right is the old *Quaker Burial Ground.* Although the Dutch were the first to establish a settlement on Tortola in 1648, a colony of Quakers from the United States followed shortly thereafter. The small white Methodist church near the village of East End at *Chapel Hill* on the right was built over 300 years ago and was originally the church that the Quakers used. It has been damaged by several hurricanes and most recently was restored in 1974. Today, scores of persons on St. John and Tortola bear the surname Penn, dating back to the Quaker activities on Tortola.

The short one-lane toll bridge (opened in 1966) spanning the narrow Beef Island Channel near the village of East End connects Tortola with *Beef Island.* This is where the *main airport* (surfaced in 1976) for the BVIs is located. However, its limited length (4,500 feet, or 1,500 meters) cannot accommodate large jet planes. As a result, there are plans to build another airport on Anegada Island to the north because it has more available level land. *Trellis Bay,* on the northern side of Beef Island, has become a favorite yacht-anchoring spot for sailors passing through this area on their way to other islands. In its center is small Bellamy Cay, site of a modest but interesting restaurant known as *The Last Resort.* Yachtsmen who are anchored in the bay frequently travel by small boats to the key to have dinner and listen to its owner, Tony Snell, perform a comedy routine and sing songs while playing his guitar. It is also accessible by a ferry service provided by the restaurant.

RIDGE ROAD, THIRTY MINUTES

After retracing the path about a mile (1.6 kilometers) back to the town of *Long Swamp,* turn left and continue west over *Ridge Road* high atop the mountainous backbone of Tortola. From this ridge look down toward the ocean and you will see *Josiah's Bay* and its small but pretty beach. Just inland from it you will also see the reddish brown Josiah's Bay Pond. The latter is a freshwater pond fed by intermittent streams (called guts in the English-speaking Caribbean) that drain downslope into it. Since the streams are usually dry, the pond tends to almost dry up during the drier winter season, resulting in its distinctive color. The Tamarind Country Club Hotel, a small resort on the side of the slope leading down to Josiah's Bay, has specifically been designed for tourists who want to get away from it all. Notice the denuded vegetation that scars the hillsides here, a result of overgrazing by goats and sheep. Also, look at what appear to be very small terraces but are really erosion trails left by the animals that grazed this area. As you proceed along Ridge Road look toward the northwest for a good view of *Jost Van Dyke*. On a clear day you also will be able to see the low silhouette of *Anegada Island* off in the distance to the north. More than 20 miles (32 kilometers) of reefs off the coast of Anegada pose a major problem to shipping in this area. At least 300 ships have wrecked on these reefs. Farther westward near Turnbull Hill and Fahie Hill you will see several small depressions created by bulldozers. Their purpose is to collect rainwater for livestock and wild animals. In the BVIs they call these ponds slobs.

BREWER'S BAY TO CANE GARDEN BAY BEACH, ONE HOUR

Just west of Great Mountain (the mountain that overlooks Road Town) is the turnoff for *Brewer's Bay*. Take the steep road (Brewer's Bay Road East) down to it. As the road winds down the mountain slope notice how the vegetation is more lush on these slopes than it was on the southern side of the islands when you were traveling along Waterfront Drive toward Beef Island. Enjoy the beautiful beach at Brewer's Bay either by a swim or by having a drink from its small refreshment stand. This is the beach the locals prefer because it is just far enough away from Road Town not to be visited by many tourists.

Take Brewer's Bay Road West southward up the steep hill to Cane Garden Bay Road and turn left (to head southeast). As you drive up Brewer's Bay Road West, observe that some of the small farms on the hillsides in this area have small terraces cut into their steep slopes. These have not been created by grazing livestock as

*Cane Garden Bay, Tortola, British Virgin Islands. Photograph by
Thomas D. Boswell.*

was the case near Josiah's Bay; local farmers rather have purpose-
fully constructed them to slow the rate of soil erosion and to catch
rainwater. Follow the Cane Garden Bay Road and the signs to
Meyers Peak where you will arrive at the restaurant called *Sky
World.* Stop for a look from the restaurant's observation platform
at the magnificent view of the islands on the other side of Drakes
Passage. You can see St. John to the far west, Peter and Norman
Islands in the center, and the smaller islands of Salt, Cooper, and
Ginger to the east. The elevation here is 1,470 feet (448 meters).

Turn back onto Ridge Road and follow it south to the turn for
Sage Mountain to the left. Notice that the main branch of this road
veers to the right and becomes Cane Garden Bay Road, but this is
not the branch you want. You turn left onto the smaller road that is
a southern extension of Ridge Road. You can drive on this dirt road

to the base of *Sage Mountain*'s summit. Once there, you can hike the rest of the way to its crest (it will take about twenty minutes). This is the highest peak (1,740 feet, or 530 meters) in the entire Virgin Islands. Turn northward and retrace some of your tracks to Chalwell, where the radio and telecommunication antenna for Tortola are located. Two telecommunication cables run down the side of the hill from here to the ocean and then continue as submarine cables more than 800 miles (1,288 kilometers) to Bermuda; they serve as a communication conduit to Europe.

Go back via Ridge Road to Cane Garden Bay Road and follow it downhill (toward the west) to the beach of the same name. This is Tortola's most-used beach and its sheer beauty tells you why. There are usually a number of yachts in the bay and many tourists on the beautiful white sand beach. Notice the medical clinic and primary school on the left. Farther along, again on the left, you will see the small rum distillery operated by the Callwood family. If it is open, it is worth visiting. Mr. Callwood grinds his own cane in a press at the back of the small brick building where he makes his rum. He pours the distilled spirits into whatever type of liquor bottle he happens to have at the time and sticks one of his labels on it. You can purchase a bottle of his product cheaply and take it home as a souvenir, but my advice is that you not drink it!

SUGAR MILL TO PORT PURCELL, FORTY-FIVE MINUTES

Follow North Coast Road south from Cane Garden Bay along the waterfront. Pass over a very winding part of the road through Windy Hill on your way to Great Carrot Bay and proceed southward to Apple Bay where you will find Sugar Mill Estate Hotel on your left. This beautiful complex of some twenty rooms was built on the site of a former sugar mill.

Continue southward on North Coast Road and follow its extension, known as Zion Hill Road, to Drake's Highway (which runs along the south shore to Road Town). Turn right (toward the west) and follow this road (it is now known as West End Road) to its end and the settlement of *West End*. The harbor, here called *Soper's Hole,* is where the ferries from St. John and St. Thomas arrive. Usually, a number of yachts are anchored here and there is a small customs house used to process goods being brought from the USVIs. Take the small bridge south to Frenchman's Cay and Pusser's Landing for a breathtaking view of the island of St. John, which is only about 2 miles (3 kilometers) to the west.

Turn back eastward along West End Road and return to Drake's Highway. Follow this road all the way back to Road Town. Just east of the small settlement of *Recovery* you will see some quarries on your left. From these sites stone is mined for use in road construction throughout the island. Continue eastward until you see *Nanny Cay* on your right. It used to be a small raised sand bank with mangroves offshore, but was joined with landfill to the mainland. It has been made into Tortola's second largest marina, with space for more than 200 yachts. Farther east near Sea Cow Bay is the village of *Albion*, home of Tortola's only horse-racing track (to your left). This facility is also used for other activities, especially during holidays. Proceeding along Drake's Highway toward Road Town, just before reaching the western edge of Road Bay, you will arrive at the largest hotel and resort complex in the BVIs, *Prospect Reef Resort*. It has 131 rooms and is a complete resort with tennis courts, two pools (one freshwater and the other saltwater), and its own harbor and marina. *Fort Burt* is located on the hill in the Fisher Estate area, above Burt Point and Careenage Hole (locally referred to as Cleaning Hole). The Dutch built it during the late 1660s to protect Road Harbour; it is now used as a hotel and restaurant.

As you continue along the road it becomes renamed Waterfront Drive and is close to where this field trip began. As the road swings northward you enter Road Town. On the left is *Government House,* the residence of the governor of the BVIs, which was built during the 1920s. A short distance away is *Peebles Hospital* (the town's sixty-bed infirmary) on your left. This is the only hospital in the BVIs, although some of its other islands have clinics. Farther along are several government offices on both the right and left sides of the road. On the left is a ferry dock used for boats that run to the other BVIs, as well as to the USVIs. Near the ferry dock and between Waterfront Drive and Mainstreet is Road Town's old market place and town center, *Sir Olva George's*

Square. As Waterfront Road bends toward the northwest, you are traveling on land reclaimed by land-filling operations that connected the mainland to a small key. This fill area is called Wickham's Cay. About 0.25 mile (0.4 kilometer) north is a second similarly reclaimed area known as Wickham's Cay 2. Here is the largest marina in the BVIs, known as *The Moorings*. It is home base to the largest yacht charter fleet in the world, with more than 100 boats based in Road Town alone and hundreds elsewhere throughout the Caribbean, Tahiti, Mexico, Hawaii, Tonga, and the Mediterranean Sea. Continue driving 0.3 mile (0.5 kilometer) to return to where this field trip started at Port Purcell.

St. Thomas, one day

Two field trips have been designed for St. Thomas. One is a driving trip around the island and the second is a walking tour of its most important city, Charlotte Amalie. Both are half-day trips. St. Thomas is about 13 miles (21 kilometers) long and 3 miles (5 kilometers) wide, encompassing approximately 32 square miles (8 square kilometers). With a population of about 50,000 it has one of the highest person/land densities in the entire Caribbean. But the true population pressure is even higher than these figure suggest because close to 2 million visitors come to the island each year by both air and sea. The type of tourism that has been developed on St. Thomas is more oriented toward short-term visitors (who stay for seven to ten days) and large masses of tourists, more so than in St. John, St. Croix, or the British Virgin Islands. As is the case with most places, the population of the island is unevenly distributed with most living in the area of Charlotte Amalie.

AROUND THE ISLAND, TWO TO FOUR HOURS

Because of the complicated road system on St. Thomas, the heavy traffic in Charlotte Amalie, and the fact that people here drive on the left side of the road, it is recommended that the visitor hire a cab to handle the driving for this trip. Depending on the number and lengths of stops and traffic, it will take between two and four

St. Thomas, United States Virgin Islands

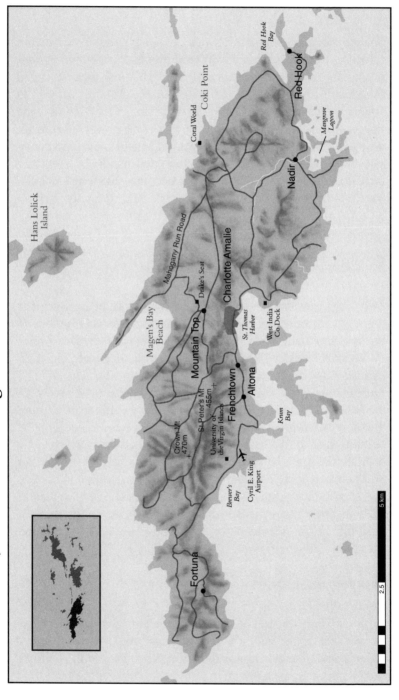

hours to complete this excursion. It will begin at the West Indian Company Dock, where most of the visiting cruise ships are berthed.

WEST INDIA COMPANY DOCK TO THE UNIVERSITY OF THE VIRGIN ISLANDS

Leave the *West India Company Dock* and proceed westward on Route 30. The name of this street changes (French Bay Road, Long Bay Road, and Veterans Drive as you drive from east to west), but you will have no problems as long as you stay on the road closest to the water. You will pass along the waterfront of Charlotte Amalie.

As you are about to leave the western end of Charlotte Amalie there is an interesting area known as *French Town* approached by a road next to the U.S. Post Office, to the south of the road at the base of Careen Hill next to Hassel Island. Take a quick spin through it and you will see that the names of most of its streets begin with the prefix *rue* (meaning street in French). This is an area that was settled by Frenchmen who came from the French Antilles island of Saint-Barthélemy (called St. Barts for short) earlier this century. There is a distinct social difference between the French of French Town and the rest of St. Thomas. Until the past thirty years, many of the residents of this area were fishermen and also farmed along some of the north-facing slopes of St. Thomas, especially near Magen's and Hull bays. French Town now is largely a residential area with a few very good restaurants.

After you return to Route 30 and head west for about 0.5 mile (0.8 kilometer) you should turn left on Route 304. Follow it for about a mile (1.6 kilometers) to the point where it intersects with the road to the airport (Route 302). Along the way you will see a National Guard Armory on your right and a marina and small cruise-ship berth on the left. Also, you will see the works of the Virgin Islands Water and Power Authority (WAPA). There is a *desalination plant* near the head of Krum Bay and this is where St. Thomas gets much of its drinking water.

UNIVERSITY OF THE VIRGIN ISLANDS TO MOUNTAIN TOP

After Route 304 intersects with Route 302, take the latter north (away from the airport). In about 0.25 mile (0.4 kilometer) it will intersect with Route 30 and you should take it in a westerly direction. Within 0.3 mile (0.5 kilometer) you will see the campus of the *University of the Virgin Islands* (UVI) on the hill to your right. Drive up to it to see this beautiful little campus, where about 5,000 students from both the USVIs and BVIs, as well as from other Caribbean islands, attend school. Once you are up there look downhill and southward to get a good view of Cyril E. King Airport, named after the USVIs first native-born and popularly elected governor. Its terminal was rebuilt in 1990 and in 1991 its runway was extended and otherwise improved.

Drive downhill from UVI and head eastward on Route 30 back into town. When you reach the center of town, find Solberg Road (Route 40) and follow it uphill. Because of the complicated maze-like nature of the small roads here, this is where you will be glad you have somebody driving for you, if you took my earlier advice. As you travel away from Charlotte Amalie and upward into the mountains there are two interesting changes you should notice. First, note the change in *socioeconomic status* of the neighbor-hoods as you move farther upslope, from the poorer residential areas near downtown to the more affluent neighborhoods the far-ther you proceed up the mountain (with better views and more expensive real estate). Second, notice how the *vegetation* changes from scrub and bush on the drier lowlands to a more luxuriant forest of larger trees and heavier foliage near the top of the moun-tain. This is a reflection of the greater precipitation that occurs at higher elevations here (30 to 40 inches, or 75 to 100 centimeters at the St. Thomas Harbor next to Charlotte Amalie and perhaps 50 to 60 inches, or 115 to 150 centimeters, at Mountain Top).

Follow Solberg Road (Route 40) and turn right (east) on Moun-tain Road (Route 33). Follow the signs (for about 0.3 mile, or 0.5 kilometer) to *Mountain Top,* near the top of St. Peter Mountain (about 1,500 feet, or 500 meters, above sea level). This is the second highest peak (after Crown Mountain which has an eleva-tion of 1,550 feet, or 515 meters) on St. Thomas. There is a restaurant, bar, and souvenir shop at Mountain Top, but the real attraction is the spectacular view of the BVIs to the north. This is where the banana daiquiri was invented.

DRAKE'S SEAT TO MANGROVE LAGOON

Go back to Mountain Road (Route 33) and follow it for about 0.3 mile (0.5 kilometer) to Skyline Drive (Route 40). Take Route 40 to *Drake's Seat* (about 0.3 mile, or 0.5 kilometer, toward the east). Stop here for the most spectacular view in St. Thomas. You can see Magen's Bay Beach to the north and the BVIs and Drake's Passage to the northeast. The pirate Sir Francis Drake is supposed to have used this as a lair for spying on the maritime traffic that passed through this area.

Follow Route 40 eastward to where it intersects Route 35 and follow the latter downhill to *Magen's Bay Beach.* Most Caribbean guidebooks tout this as one of the most beautiful beaches in the world. Its horseshoe bay is lined with powdery white sand and turquoise water. Also, on the grounds of the park is a coconut plantation.

Take Route 35 uphill to where it joins Route 42 (*Mahogany Run Road*) and follow the latter eastward. Follow this road with luxuriant vegetation and beautiful views for about 4 miles (6 kilometers) to Smith Bay Road (Route 38) and head east-northeast to the turnoff (about 1 mile, or 1.6 kilometers) for *Coral World,* on Coki Point. This is one of the world's famous sea aquariums. It has a three-story observation tank about 50 yards (46 meters) offshore in the ocean that is used to view the natural marine coral environment of this area. Also, it has a reef tank that simulates a submarine view of an artificial reef. In addition, a number of its other tanks contain sea life.

Backtrack on the road from Coral World to Smith Bay Road (Route 38) and follow the latter eastward to *Redhook Bay.* This is one of the locations for the ferries that run to St. John and the USVIs. Also here are the main office for the Virgin Islands National Park Service and a marina.

Route 38 becomes Route 32 (Redhook Road) as this road is followed to the southwest. Follow it westward to where it intersects Route 30. Along the way you will see a number of small beach resorts on the seaward side of Redhook Road and several condominium complexes. If you detour off Route 32 onto Route 322 and head toward Cowpet Bay, you will see the newest and largest area of condominium development on the island. Also, near the town of Nadir (back on Route 32) you see the only horse racetrack on St. Thomas.

Follow Route 30 back to the West India Company Dock. You pass along the edge of a mangrove swamp at Mangrove Lagoon. This route takes you through another area being developed for condominium sales. Also, there are some beach resorts on the seaward side of the road, but to get to most of these you have to turn and follow small roads that isolate them from the main part of Route 30. Among these are such resorts as Bolongo Bay, Limetree Beach, Morning Star Beach Club, and Frenchman's Reef Hotel Resort, and some of the most expensive real estate in the Virgin Islands.

CHARLOTTE AMALIE, TWO TO THREE HOURS

When Denmark first settled what is today Charlotte Amalie it was named Tap Hus or House of Drink. But in 1691 it was given its present name in honor of a Danish queen. Although the Danes were first interested in a plantation economy based upon the production of sugarcane, when the Danish Crown gave Charlotte Amalie free port status in 1764, shipping and commerce soon replaced sugar in value. This was partly because St. Thomas's steep slopes and unreliable rainfall were never well suited to the production of sugar. But in addition to these unattractive qualities, it is equally clear that the port of Charlotte Amalie had some definite advantages that allowed it to become "the emporium of the West Indies" by the end of the eighteenth century. For instance, it had a *strategic location,* close to the main travel lanes that connected Europe and the New World. During the U.S. Civil War ships were frequently outfitted here to run the Union blockade. Also, Denmark's *neutrality* and *free port status* gave this port an advantage over those belonging to warring nations. In addition, Charlotte Amalie has one of the *finest natural harbors* in the West Indies because of its size, protection, and depth, primarily a result of its being located in the collapsed crater (called a caldera) of a long-extinct volcano.

Many of the buildings in Charlotte Amalie date from the early nineteenth century, after a couple of disastrous fires destroyed most of what had previously been there. According to the U.S. Department of the Interior, there are more 100-year-old buildings still in use in this town per unit area than any other city in the United States.

Within its legal limits, the city is less than 2 miles (3 kilometers) long and about 0.5 mile (0.8 kilometer) wide. There are three major east–west running streets: Veterans Drive along the waterfront (with many shops and eating establishments); Dronningens Gade (now usually called Main Street), which is the major shopping street; and Wimmelskafts Gade (now Back Street), another shopping street. These three streets are connected by a number of short streets (called *gades*) and alleyways, which also have a plethora of retail shops. Because distances are short and traffic is congested in downtown Charlotte Amalie it is best to follow the itinerary of this field trip on foot.

This field trip also begins at the West Indian Dock. Immediately across the street from the dock toward the southeast is Havensight Mall, where it is possible to find some of the items sold downtown. Follow Route 30 (called Long Bay Road here) into town, walking toward the west. On your right, the first glimpse you will have of the city is not its most flattering view because this is where the island's oldest low-income housing project is located (called *Paul M. Pearson Garden*). It was built in 1953 and is shows its age. Just to the east of it is another project called *Oswald Harris Court,* constructed in 1962. Approximately one fifth of the USVI's population lives in housing projects like these, built by the U.S. Government and the USVI Housing Authority. Inland and upslope from Pearson Garden is the only hospital in St. Thomas.

As you pass around Frederiksberg Point and head into town look up on the hill above you to see *Bluebeard's Castle*. It was built during the late 1600s but has been refurbished and now serves as an elegant hotel and restaurant. It is worth a walk up there for the outstanding view of the city below.

Back on Veterans Drive and walking westward you will pass by the multi-storied Federal Office Building and the main police station of St. Thomas on your right. Next you will see a large dark red building on your right that is *Fort Christian*. It was originally erected in 1676 to protect the early colonists from pirates and foreign intruders. Now it houses the Virgin Islands Museum and a fire station, but it has also served as a church, governor's residence, police station, and prison. It is the oldest building in continuous use in the USVI.

Across the street to the south of Fort Christian is a two-story green building, USVI's *Legislature Building*. It was constructed in 1874 and was a Danish marine barracks until the United States purchased the Danish Virgin Islands in 1917. Since 1957 it has housed the USVI's Senate.

Inland from the parking lot, immediately west of Fort Christian is *Emancipation Gardens Park*. This where the freedom of the slaves was proclaimed in 1848. Next door on the corner of Tolbod Gade and Norre Gade is the *Grand Hotel*. It was built in 1841 as the Commercial Hotel. Now it houses shops and offices. It is a marvelous representation of the way hotels were built in the Caribbean during the middle 1800s.

On the northern side of Norre Gade is *Frederick Lutheran Evangelical Church*. Built in 1820, it is the second oldest Lutheran church in the Western Hemisphere. Walk up to Kongen's Gade one street above Norre Gade and you are in the Government Hill district. On the right is *Government House*. It is the official residence of the USVIs governor. It was built by the Danes during the mid-1860s. Next to Government House are the famous *99 Steps*. Before use of automobiles, steps were used to climb the steep hills behind Charlotte Amalie, as in San Juan, Puerto Rico. Just west of the 99 Steps is *Hotel 1829,* built in 1829. It is another good example of nineteenth-century colonial architecture. Originally it served as the residence of a Danish sea captain named Lavalette.

Above the 99 Steps is *Blackbeard's Castle,* built in 1679. The old stone tower is now used as a hotel, similar to Bluebeard's Castle. Walk west to Crystal Gade. On its southeastern corner with Nye Gade is the *Dutch Reformed Church*. This church has the oldest congregation on St. Thomas, which dates to 1660, although the building was erected in 1844. Continue about one block westward on Crystal Gade and you will come to *St. Thomas Synago-gue*. It is the Western Hemisphere's second oldest synagogue (only the one in Curaçao is older) and maintains the Sephardic tradition of having clean sand on the floor that is swept to commemorate the flight of the Jews out of the Egyptian desert. Its first version was built in 1796, but it was destroyed by fire and the current edition was built during the early 1800s. On Berg Hill, upslope from the synagogue, are the residence and offices of the *Danish Consulate*. This large white building is recognizable by the red and white flags that fly over it. It was built as a private home by a former Danish governor.

The rest of the downtown area of Charlotte Amalie is worth walking around and taking pictures of not only because of its shopping opportunities. Most of the buildings that house the shops along the gades between Veterans Drive, Main Street, and Back Street were old warehouses and other buildings used by the Danish traders during the early 1800s and are thus close to 200 years old. One interesting sight is *Market Square,* west between Main and Back Streets next to Strand Gade. It has been here for more than 200 years and even today the few remaining farmers living on St. Thomas bring their agricultural produce here to be sold. Because of its central location and the scarcity of parking places elsewhere, taxi cabs can be found here during the evening waiting for calls for their services over their radios.

St. John, three hours

St. John is the smallest of the three main U.S. Virgin Islands, with an east–west extent of about 9 miles (15 kilometers) and north–south width of approximately 3 miles (5 kilometers). The total area is only 20 square miles (53 square kilometers). However, what makes St. John exceptional is not its size but the manner in which development and growth have been controlled.

In 1952 Laurance Rockefeller purchased the Caneel Bay Planta-tion on the northwestern end of St. John and developed it into a resort complex built in harmony with its physical environment.

St. John, United States Virgin Islands

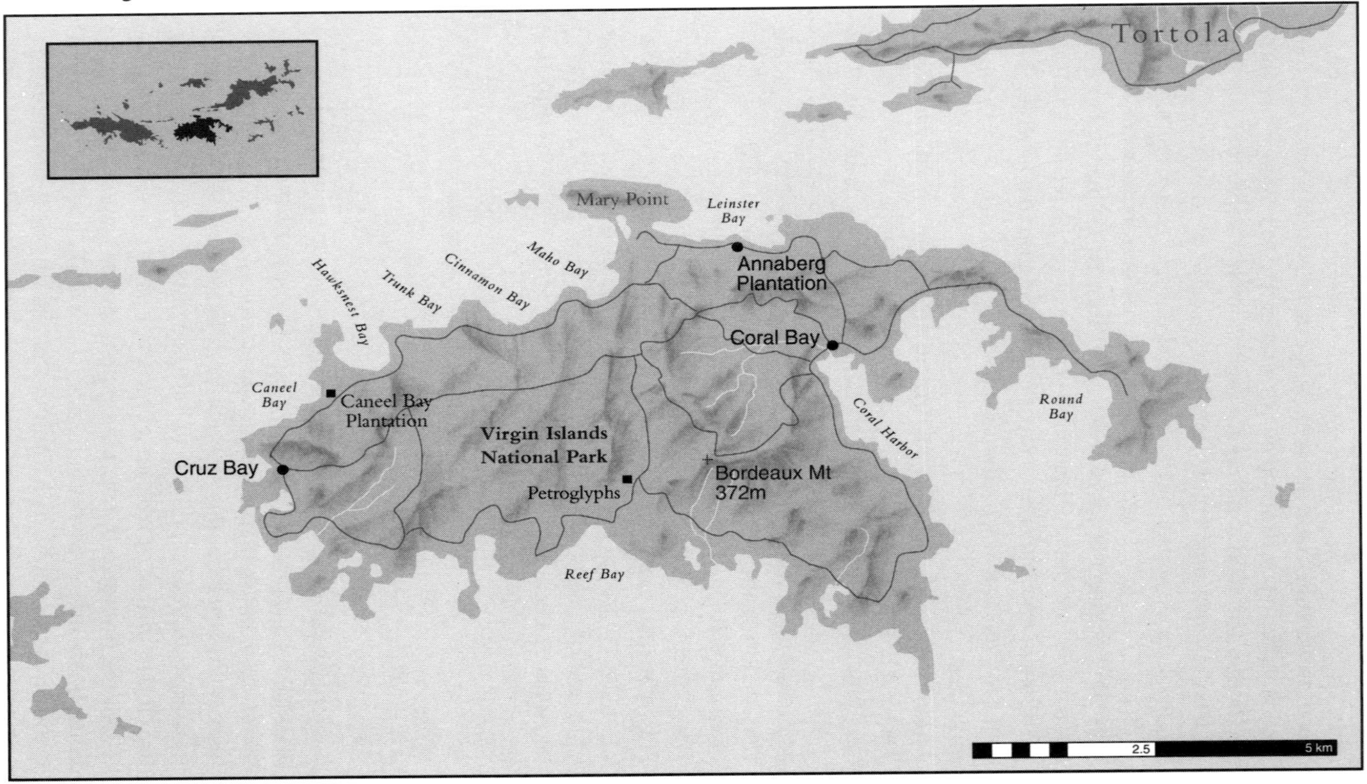

Tortola

Round Bay

Coral Harbor

Leinster Bay

Annaberg Plantation

Coral Bay

Mary Point

Bordeaux Mt 372m

Maho Bay

Cinnamon Bay

Virgin Islands National Park

Petroglyphs

Reef Bay

Trunk Bay

Hawksnest Bay

Caneel Bay Plantation

Caneel Bay

Cruz Bay

5 km

2.5

Then between 1955 and 1956 he purchased an additional 5,000 acres (2,024 hectares) which he donated for use as a park to the U.S. National Park Service. The park became known, when it opened in 1956, as the *Virgin Islands National Park*. In 1984 the Rockefellers further agreed to donate most of the rest of the Caneel Bay land to the national park by the year 2025. Today, almost seventy percent of the island is included within the park. Development projects inside its confines are strictly controlled by the federal government. As a consequence, most of the park has been left in a natural state and is expected to remain so into the foreseeable future.

This field trip will emphasize the controlled development and natural environment of St. John. It starts at the town of Cruz Bay on the island's western end. Perhaps three fourths of the island's 4,000 residents live in this settlement. In short, it is the economic and cultural epicenter of St. John. But when the Danish first settled the island in 1716 the village of *Coral Bay* on the eastern coast was the island's primary settlement. By 1800 the development on St. Thomas and growth of Charlotte Amalie as a major trading emporium redirected growth to the western side of St. John, and Cruz Bay with its excellent harbor. Coral Harbor is a far superior natural embayment than Cruz Bay and, unlike the latter, could accommodate large cruise ships. Some reports suggest that the Danes moved from Coral (Kraal) Bay to Cruz Bay to escape sandfly infestations. Nevertheless, today, the ferry from St. Thomas arrives in Cruz Bay and so does virtually everything else. The island's three main roads also begin here and so will this excursion.

CRUZ BAY TO PETROGLYPHS, AN HOUR AND A HALF

The small town of Cruz Bay is an uncomplicated settlement. Virtually all its commercial activity takes place within a three-block radius of the ferry dock. One block to the north and overlooking the town's bayfront park from a small hill is its most historic building, *The Battery*. It was constructed in 1735 to guard the town against further slave revolts, after the insurrection in 1733, the first such revolt anywhere in the Western Hemisphere. It now houses the administrative offices of the government of St. John and is also called *Government House*. Northeastward, across another small embayment from The Battery, you will find the Visitor's Center for the Virgin Islands National Park.

From the terminal where the ferries dock, drive straight ahead up the street called Centerline Road (Route 10). As you proceed upslope toward the east you will pass through a dry area where the annual precipitation is about 40 inches (100 centimeters) and scrub and bushes prevail. This considerably eroded area has been grazed by goats. About a mile (1.6 kilometers) along the road next to Margaret Hill you can look back for a panoramic view of Cruz Bay. About a mile farther along you pass the turn-off for the island's garbage dump, which has become a center of ecological controversy because of the ground pollution it has created.

Continue eastward for approximately another 2 miles (3 kilometers) to the walking trail that leads southward to *Reef Bay*. Notice that at this higher elevation the vegetation has become greener and more trees are growing, reflecting the greater precipitation here. If you follow this trail downslope you will see a dramatic change in the vegetation to dryer conditions, like those in the vicinity of Cruz Bay. The National Park Service organizes a 2.5-mile (4-kilometer) field trip that follows this trail to the sea. Along the way the area's history of former sugarcane production is discussed. Remains of four sugar estates, and more recently abandoned farming communities, can be seen along the way. About 0.5 mile (0.8 kilometer) from the end of the trail toward the sea is another trail that leads about one quarter of a mile toward the west to the site of some ancient stone carvings, known as *petroglyphs*. Exactly who carved them is still debated. Some suggest that pre-Columbian Indians (perhaps descents of the Arawaks) carved them. Others think newly arrived African slaves carved them during the late 1600s. One epigrapher, Dr. Barry Fell, thinks pre-Columbian Africans made them. (This information was provided through a conversation with former St. John schoolteacher and naturalist Doris Jadan, Cruz Bay, St. John, on 2 May 1991.)

BORDEAUX MOUNTAIN AND RETURN TO CRUZ BAY, AN HOUR AND A HALF

Continue eastward along Centerline Road until it joins Bordeaux Mountain Road (Route 108). Follow the latter and it will pass near the summit of *Bordeaux Mountain,* the highest elevation in St. John (1,277 feet or 390 meters). Continue along this road and follow it to Coral Harbor, which some think is the finest natural harbor in the West Indies. It was here that the Danes built their original community on St. John.

Follow East End Road (Route 10) out of Coral Bay to the east for about 3.5 miles (5.6 kilometers) to Haulover Bay on the peninsula of East End. Here the land narrows to a little more than 100 yards (91 meters) and you can see water on both the north and south sides. The vegetation is a reflection of almost arid conditions here. To the north is a spectacular view of Tortola in the BVIs, only about 2 miles (3 kilometers) away.

Return to Coral Bay via East End Road and follow Kings Hill Road (Route 20) about 1.5 miles (2.4 kilometers) to Centerline Road. Cross the latter and Route 20 now becomes known as North Shore Road. Follow it for about half a mile and take the turnoff to the right to *Annaberg Plantation,* a property that dates from 1721. A short hike of about one-quarter mile will take you to one of the most famous plantation ruins in the West Indies. The National Parks Service offers guided tours and produced a pamphlet that provides directions for a self-guiding tour of this facility. The ruins lie on a hill that overlooks Leinster Bay, and from which there is a good view of the BVIs to the north and the east. You can also see the peninsula of Mary Point. From the cliffs on this point, a number of slaves who revolted against their Danish plantation masters in 1733 to 1734 are believed to have committed suicide by jumping to the rocks below. Other slaves shot themselves on the hill above Ram's Head on the extreme southeastern tip of St. John. All West Indian islands have stories of slaves committing suicide because of their harsh treatment. It was popularly believed that the red color of the rocks came from the bloodstains of the Indians who jumped to their deaths. In fact, the color is derived from the mineral hematite contained in these rocks. In any event, this slave rebellion was quelled by French troops from Martinique who had come to the aid of the Danes.

Proceed back to Route 20 and continue westward. Along the way you will see some of the most beautiful beaches in the Caribbean. Among these are Maho Bay, Cinnamon Bay, Trunk Bay, and Hawksnest Bay. The first two have campgrounds and Trunk Bay has a submarine trail marked on its coral reef. When you reach Caneel Bay, drive into the grounds and look around. This is the old plantation facility (started during the 1700s) that Laurance Rockefeller purchased in 1952. The Caneel Bay complex is a complete resort run by Rock Resorts, Inc., which is scheduled to turn over the grounds to the Virgin Islands National Park in the year 2025. The complex has been built around the ruins of the old sugar mill and distillery. It employs about 450 persons and provides facilities for about 350 guests. There are three restaurants, seven beaches, and eleven tennis courts here.

Drive westward over North Shore Road back to Cruz Bay. Along the way there will be many beautiful vistas overlooking the water. This field trip ends at Cruz Bay.

△ *Day Three*

SINT MAARTEN AND
SAINT-MARTIN—A DIVIDED ISLAND

Perhaps the most interesting aspect of this island is its name, which in turn is a consequence of its unique history. Of the 37 square miles (about 96 square kilometers) its southern forty-three percent is owned by the Netherlands, while its northern fifty-seven percent is a possession of France. The Dutch call their part Sint Maarten (popularly spelled St. Maarten) and the French call theirs Saint-Martin (shortened to St. Martin). This is the smallest place in the world that is divided between two sovereign powers.

Although the Spanish were the first to claim the island during the second voyage of Columbus in 1493, the Dutch were the first to settle it during the 1620s, when they became interested in its several shallow ponds as a supply of salt for the Dutch herring industry. The Spanish recaptured the island in 1633 but soon they once again abandoned it to concentrate their efforts on the more prosperous Greater Antilles. Shortly afterward a few Dutch and French settlers occupied the island and, in a rare act of diplomacy, signed an agreement in 1648 to split the jurisdiction of the territory as it is now. But contrary to popular belief this did not end the controversy over ownership. In fact, the island was destined to undergo sixteen changes in ownership among France, the Netherlands, and England. Finally, in 1816, the 1648 treaty signed between the French and Dutch was again ratified. Since then peace has reigned between the two sides of the island and today there is free passage between the two jurisdictions. A monument about

St. Martin / Sint Maarten

Grand Case Bay

Grand Case

French Cul-de-Sac

Orient Beach

Baie de l'Embouchure

Etang aux Poissons

Paradis Peak
+ 424m

Quartier
d'Orleans

France
Netherlands

Oyster
Pond

Quartier du
Colombier

Marigot

Great Salt
Pond

Philipsburg

Great
Bay

Fort
Amsterdam

Pointe
Du Bluff

Baie de Nettle

Simpson Bay
Lagoon

Cole
Bay

Baie
Rouge

Simpson Bay

Mullet Pond Bay

10 km   5

halfway along the road between Philipsburg, capital of the Dutch side, and Marigot, capital of French side, is the only indication that a boundary exists.

The Dutch side has a slightly larger population (27,000) than the French side (25,000). It is also somewhat more developed, but the French side is catching up. First the salt industry and later plantation agriculture (emphasizing the production of sugar and to a lesser extent tobacco and cotton) served as the economic mainstays of the island. Since the 1970s, however, tourism has reigned supreme on the Dutch side and became important on the French side about a decade later. Today, duty-free shopping, casinos, a wide range of accommodations, beautiful beaches, a first-class airport, and the fact that it is a good "jumping-off" place for visiting nearby islands enhance its appeal to tourists.

Although the island was formed 50 million to 60 million years ago by volcanic activity, it is no longer tectonicly active. The highest summit in its hilly landscape (Paradise Peak) is only 1,391 feet (424 meters) above sea level. In spite of its average annual rainfall of about 45 inches, the vegetation exhibits a decidedly arid character. This is because the hills are not high enough to produce more rain and the steep slopes and shallow soils inhibit moisture retention and increase runoff to the numerous salt ponds and surrounding ocean.

Although Dutch and French are the official languages of their respective sides of the island, in fact almost everyone on the Dutch side speaks English fluently and most people on the French side speak some English. St. Maarten is now governed as part of the Netherlands Antilles, which also includes the four islands of Saba, St. Eustatius, Curaçao, and Bonaire. All its permanent residents are citizens of the Netherlands. St. Martin is similarly governed as a part of the Département of Guadeloupe and all its citizens are citizens of France.

Sint Maarten, forty-five minutes

This excursion begins at *Little Pier* in Philipsburg. If you came by cruise ship, this is where you will arrive via small boat, while the ship remains anchored farther out in the harbor. Philipsburg is located on an old sandbar between Great Bay and Great Salt Pond. It has two major streets, Front Street (along the shore of Great Bay) and Back Street. During the early 1970s additional land was reclaimed from Great Salt Pond and now there are two (in its western part) and three (on its eastern half) new streets. Residents usually call this new area "Pond Fill."

Most tourists shop on Front Street and the small alleyways that connect it to Back Street. The alleys have Dutch names that usually end with the suffix *steeg*, which means alley. Front Street and some of these connecting alleys resemble a shopping mall, especially when several cruise ships are anchored in the bay. Back Street and the other streets that parallel it contain more modest and less expensive retail and service enterprises that cater more to the local population than to tourists.

Because of the growth of hotels, casinos, restaurants, and retail shops, few buildings of historical significance are left in Philipsburg. One exception is the old *Courthouse,* built in 1793. Its upper floor is still used for court proceedings and its ground floor is used as the town's main post office. East on Front Street are two other historical houses, they were built during the 1800s. One is the *West Indian Tavern,* which is an excellent example of the type of buildings built more than a century ago with recently added West Indian ornate embellishments. Across the street is the *Pasanggrahan Hotel.* It was originally known as the *Royal Guest House* because it used to provide space for such luminaries as former Queen Wilhelmina of the Netherlands. The name Pasanggrahan means guesthouse in Indonesian, reflecting the ties that the Dutch had with their former Southeast Asian colony.

A small peninsula pokes its head southward as the southwestern side of Great Bay. Here a hill contains St. Maarten's most historic site, *Fort Amsterdam.* The Dutch built this fort in 1631 and have added to it several times since then. This was the first Dutch fort

Salt Pond, St. Barthelemy, French Antilles. Photograph by Thomas D. Boswell.

built in the Caribbean and was constructed to defend Great Salt Pond before Philipsburg was built. It was here that one of the founding fathers of New York City (then New Amsterdam), Peter Stuyvesant, lost one of his legs in a battle with the Spanish. The fort can be reached by parking your car at the Divi Little Bay Beach Resort and hiking about five minutes upslope. In the nineteenth century this old fort ceased military operations and until the 1950s it was used as a signaling station and for radio communications. Since 1970 the peninsula on which the fort is located has been sold several times. It is currently owned by a hotel company that is trying to develop a hotel complex around it. On a nearby hill is another fort, *Fort Willem,* which the British built in 1801 and to which the Dutch added in 1816. It is now dilapidated and is not as impressive as Fort Amsterdam. Its location is easily recognizable by the television transmitting tower on the summit of the same hill. This fort can best be reached by parking next to Great Bay Resort Hotel and hiking about forty-five minutes uphill to it. Once you have made the climb, the views you will have of Little Bay and Great Bay make the effort well worth while.

As you leave *Little Pier* at the beginning of this trip you will pass immediately into *Wathey Square*. On your right-hand side is a small tourism information office and to the left is a taxi stand where you can make arrangements to hire a cab or van for your trip. As you drive northward you will pass the old Courthouse and Post Office and you should continue past Back Street to the farthest of the Pond Fill roads, called Walter Nisbeth Road. Turn right here but look to your left and you will see a small neck of land that protrudes into Great Salt Pond. It too has been reclaimed and on it is the *University of St. Martin*. This is a private school that is affiliated with Johnston and Wales University in the United States. Students undertake their first year of study here and then transfer to the main campus in the United States. It also offers short courses to people who cannot enroll full-time.

Follow Nisbeth Road for a short distance eastward until it intersects with the road that runs along the eastern periphery of Great Salt Pond and turn left onto it. After traveling about 0.25 mile (0.4 kilometer), turn left on The Arch Road. You will come to the old saltworks used around the turn of the century. Most of the salt derived from Great Salt Pond was obtained by traditional fresh-air evaporation methods. But owners of these works also experimented with a new technique where brine was pumped into tanks. The tanks were then heated with fires in an attempt to speed evaporation. In the end this process proved to be uneconomical and the works were closed down. You should notice the ecological disaster in this area. It is being turned into a garbage dump. All of this land has been reclaimed from the pond and inevitably more will be similarly reclaimed. A good estimate is that already about one fourth of Great Salt Pond has been filled in. Toward the north edge of the pond you will see a stone dike that separates the pond from fresh water that has drained toward it from the nearby hills.

Drive westward along the road that runs by the northern periphery of Great Salt Pond. You will come to the new St. Maarten Zoo. All this area along the northern shore also is landfill. Several shantytown settlements are developing in this area. They are occupied primarily by illegal immigrants from the Dominican Republic and Haiti, but also by some people who have arrived from the Lesser Antilles to the south of St. Maarten. Notice the aridity of this lowland area, as indicated by the xerophytic (drought-tolerant) vegetation, especially cacti.

At the northwestern corner of Great Salt Pond, turn right on a small road that leads northward toward the border with the French side. You will travel through an area known as the Dutch Quarter. Just after you make this turn you will pass by the *Amsterdam Shopping Center* on your right. This was built during the mid-1980s in the hope that it would attract tourists. Notice that its architecture reflects the Dutch style found on Curaçao, but found less often on St. Maarten. Unfortunately, not many tourists shop here because it is so far from Little Pier. It is too far away to walk and the expense of cab fare makes it more costly than Front Street. As a consequence it has not been very successful economically and it has never been a serious competitor for Front Street enterprises.

As you continue northward along the road you will pass through a lower-middle class district of the Dutch Quarter. When the Dutch slaves were emancipated during the latter half of the nineteenth century, they were frequently given land to live on. Much of the land in this area was acquired this way by the families now living here. As the number of illegal immigrants from both the Dominican Republic and Haiti has increased, many of the people living in this area have built shacks on their property that they rent as cheap housing to the illegal aliens. If you look closely you will see on many properties a number of these small structures and usually one larger house where the landholding family resides.

After traveling about 1 mile (1.6 kilometers), take the dirt road to the north. As this road bends east and parallels the French border, you will see one of two remaining plantation estates on the Dutch side. This was *Belvedére*, an old sugar plantation, that has been unused for almost a century, except for some low-intensity grazing livestock (first cattle and now mainly goats). The National Parks Foundation of St. Maarten is trying to convince the government of the Netherlands to purchase this land and have it set aside as a historic park. Some of the land might also be used for a low-income government housing project, to help solve some of the shantytown problem that has recently developed near here.

As you continue eastward along the road you will come to *Oyster Pond,* which now has a number of small boats and yachts in its waters. This is a well-sheltered bay, often used during hurricanes as a refuge for small vessels.

Saint-Martin, two hours

Once you have passed Oyster Pond traveling in a northerly direction, you have entered the French side of St. Martin. Notice that no signs indicate this, symbolizing the ease of movement between the two sides of the island. Next you will see several medium-sized tourist developments just north of *Oyster Pond* and at *Coralita Beach.* These hotels operate with very low occupancy rates (about ten to twenty percent of capacity). It has been said that these complexes were built with "tax dollars." The French government has offered attractive tax advantages for French citizens to invest in development projects in its Caribbean possessions. As a consequence, projects will sometimes be undertaken that do not represent good economic investments by themselves but, with the tax breaks, it may be worth the loss.

To the north of Coralita Beach is *L'Embouchure Bay* (sometimes also called Coconut Grove). On its northern end is the Le Gallion Hotel complex. The current owner wants to expand its facilities but conservationists who do not want to see this area's ecology damaged are opposing the expansion. Behind the beach (which is a sand spit) on its western side is a large lagoon, called *Étang aux Poissons.* Some illegal building has taken place here on land reclaimed from the lagoon's mangroves. You can see these structures on the right, between the road and the lagoon.

When the dirt road you have been traveling on joins a paved road, turn to the right on it and you will arrive in an area known as *The French Quarter* or *Orleans.* This was the first capital of French St. Martin. Its advantage was that it was near the most important salt ponds the French worked then. But it was not easily protected against invaders. Also, the original settlers thought that this side of the island would get more rain than it does and they hoped to develop sugar plantations here. In fact, slightly more precipitation falls on the western side of the island. As a result, in 1768 the capital was moved to Marigot. Orleans is now one of the poorer parts of the French side. It is where many of the illegal immigrants who have come to French St. Martin live in shanty-towns, similar to those on the Dutch side.

As you continue approximately 1 mile (1.6 kilometers) farther north you will come to the turnoff (to the right) for *Orient Beach.* This is the main nudist beach on the French side. There is no such beach on the Dutch side. It used to be a cotton plantation, and a few cotton bushes still grow wild in this area.

Northwest of Orient Beach is an area known as *French Cul-de-Sac,* developed since 1985 for condominiums and housing. Much has been built with the tax dollars and some of this housing also is not being used.

As the road curves west, just before arriving in the town of Grand Case, you pass the main airport on French St. Martin, called *Esperance Airport.* It was built in the middle of the salt pond, inland from Grand Case. It was expanded during the 1970s for military reasons, but it is still not large enough for large jet planes. But it can handle the medium-sized turboprops flown by Air Guadeloupe. By the way, the French word Esperance prophetically means hope in English.

As you enter *Grand Case* turn left on the main road that runs through this town. This is a charming village that has become known for its many fine French restaurants, small hotels, a few guest houses, and its beach.

About 2 miles (3 kilometers) south of Grand Case is the turnoff (to the left) for *Colombier.* This settlement is strung out along a little valley that leads up one of the spurs of Paradise Peak. It is one of the very few agricultural locations in St. Martin because there is a little more rainfall here than elsewhere on the island. A few beef and dairy cattle graze here and some food crops are grown. But, in truth, most of the agriculture is of the kitchen-garden variety for domestic use. These are not major commercial ventures. Notice the stone fences in this area. Most of these were created during the days when this land was used for plantations and they indicated property lines. They are composed of stones that were picked out of the fields so they could be farmed more easily.

As you approach the northern outskirts of Marigot you will see on your right French St. Martin's *desalination plant,* which provides most of the drinking water for the French side. Although some houses on both the Dutch and French sectors of the island have water-catching roofs and cisterns, many do not. Very little groundwater is left on the island, and much that remains is contaminated by salt water.

Although *Marigot* is the capital of French St. Martin, it does not dominate its territory like Philipsburg does the Dutch side. It is a much more subdued city, although it can be active on Saturday mornings when the central farmers' market is busiest (most of what is sold here is brought in from other islands). *Fort St. Louis* sits on the northern hill above the city. It was built in 1786 to protect the new capital when it was moved here from Orleans. The view of Margiot and the Lowlands is spectacular from the fort. Drive west on Rue de la Republique to the pier where the ferry leaves for Anguilla. On the way, on the right-hand side is the tourist bureau where you can stop for information if you need it. Marigot has duty-free shops and fine restaurants like Philipsburg, but it does not have any large hotels and casinos.

Driving down its streets is a delightful experience because many of the houses are picturesquely built in typical two-storied French West Indian style. The top floor is almost always constructed with wood and contains a porch that overhangs the front of the first floor. The bottom story is often constructed with stone or cement and usually has wooden shutters. The roofs are often made of corrugated aluminum or zinc. Drive along any of the three narrow inland streets that parallel Rue de la Liberté, which runs along the waterfront, to see these features.

Marigot, St. Martin, French Antilles. Photograph by Thomas D. Boswell.

Follow Rue de la Liberté southward out of Marigot to the *Lowlands*. The term Lowlands describes the sandy land that surrounds Simpson Bay Lagoon. As you pass over the small bridge to the Lowlands you will see some uncontrolled development and a considerable amount of poor housing curiously mixed with hotel resorts, restaurants, and nightclubs. On the peninsula to the north, *Pointe du Bluff,* is the controversial La Belle Creole resort. It was started in 1964 but ran into financial problems. It lay vacant for more than a dozen years, but was renovated and now is considered to be one of the premier hotels on the entire island. By the time you reach Baie Rouge you will notice that you are passing through a transition zone in terms of land use. You have left the uncontrolled development behind and are entering an area of expensive single-family residences called Terres Basses and Grand Etang. Most of the people who have purchased homes in this area are Americans, Canadians, and some Europeans (largely from France).

Church with mountain scenery in the background, Saba, Netherlands Antilles. Photograph by Thomas D. Boswell.

Return to Sint Maarten, fifteen minutes

Now you are entering the Dutch side of the island again. You will notice a strip of intensive resort and shopping activity in the vicinity of Mullet Bay and Maho Bay. As you continue eastward you pass the *Princess Juliana Airport,* where most international flights arrive. As you cross the small bridge you are now back on the mainland of Dutch St. Maarten. Follow this road up *Cole Bay Hill.* (Note: If you turn off to the left you will be heading back to Marigot and after about 1 mile, or 1.6 kilometers, along the way you can see the plaque commemorating the 1648 signing of the partition of the Dutch and French side of the island.) At the top of Cole Bay Hill is a lookout turnoff where you can have a great view of the Lowlands and Cole Bay. Below you in Cole Bay is Dutch St. Maarten's desalination plant and electrical power station.

As you continue back to Philipsburg follow the road to *Front Street* and drive along it (it is a one-way street where traffic moves in an easterly direction only) back to Wathey Square and the Little Pier.

St. Martin/St. Maarten is a good place to initiate travel to several other islands located in the northern sector of the Lesser Antilles. Regular travel by either boat or air is scheduled to Anguilla, St. Barts, Saba, St. Eustatius, St. Kitts, Nevis, Guadaloupe and Martinique.

Antigua

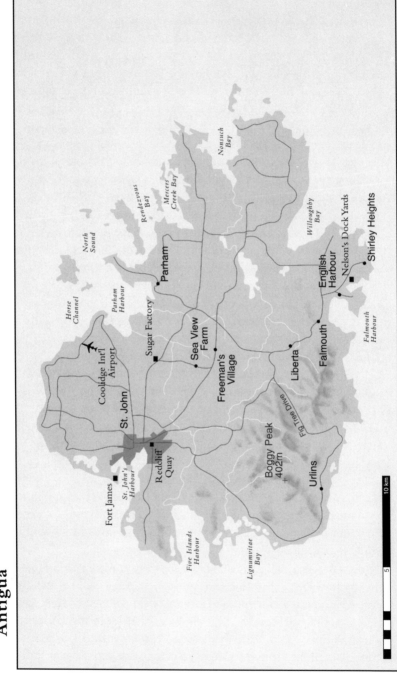

10 km
5

△ Day Four

ANTIGUA—ISLAND IN THE THROES OF CHANGE

As recently as three decades ago Antigua was a typically poor Caribbean island that relied primarily on the production of sugar for its existence. The sugar industry here had fallen on the same hard times that it had elsewhere in this part of the world. But for many workers there were few alternatives to being employed on the sugar plantations and, as a consequence, this is where almost half the labor force worked, despite low wages and underdeveloped living conditions.

Then, during the 1960s, the government began to promote tourism as means of upgrading the standard of living. Now employment in agriculture is less than one fourth of what it was in 1960 and sugarcane production has almost completely ceased. Current agriculture efforts involve growing a few vegetables, fruits, pineapples, and a little cotton and grazing livestock (mainly chickens, goats, and a few beef cattle). More than eighty percent of all food consumed on the island now must be imported, including even sugar, and less than ten percent of its income is provided through agricultural employment.

Today about sixty percent of Antigua's income is generated directly by tourism. In addition, there is a small but growing offshore banking industry and efforts are being made to attract foreign capital for development of light manufacturing. A new industrial park was constructed during the 1980s near the island's airport. Small factories produce such items as garments, paint,

Small-scale enterprise in the informal sector, Antigua. Photograph by Thomas D. Boswell.

mattresses, and galvanized metal sheets, and there is a small electronics assembly plant. Manufacturing accounts for approximately fifteen percent of the island's earnings, more than agriculture.

Still, Antigua is an underdeveloped country when compared with such countries as those of Western Europe and North America. Many people live in modest housing without inside plumbing. Pit latrines and public water standpipes are commonplace in the poorer neighborhoods of its largest town, St. John's. The economic standard of living here is certainly below that of some of the other Caribbean islands such as Puerto Rico, the U.S. Virgin Islands, Tortola, St. Martin, the Bahamas, Trinidad, and Barbados. But, on the other hand, it is higher than in most other parts of the Third World. For instance, the per capita gross national product in Antigua is between four and five times that of the Third World as a whole and is somewhat above average for the Caribbean. In general, Antiguans are better fed, housed, and medically cared for than most who live in Jamaica, the Dominican Republic, Haiti, and most of the Leeward and Windward islands formerly included in the British Empire. It is an island that is going through the throes of change from being very poor and traditional toward economic growth and modernization, with all the social adjustments and uncertainties that entails.

History

Historians still debate whether Columbus actually set foot on Antigua, but it is certain that he saw the island during his second voyage in 1493 and named it *Santa Maria de la Antigua* after a saint who resided in the cathedral of Seville, Spain. Later the name was shortened to simply *Antigua* (pronounced Anteega). Both the Spanish and French tried to settle it, but failed because of its dryness and fights with Carib Indians. In 1632, the British were the first to successfully settle it and since that time it remained under British control (except for a one-year interlude in 1666 when the French captured it) until its independence in 1981. This is one of the very few examples in the Lesser Antilles of an island that was not passed back and forth several times among the European superpowers during the 1600s and 1700s.

At first the new colonists tried growing tobacco, but in 1674 a successful planter from Barbados, named Christopher Codrington, initiated the island into the production of sugarcane. Within a decade sugar had taken over the island's economy. In 1685 Codrington leased the nearby island of Barbuda (30 miles, or 48 kilometers, to the north) from the English Crown so he could grow food for the slaves who worked on his plantation on Antigua. Since that time Barbuda has remained politically united with Antigua, as it is today. By the early 1700s more than 170 sugar mills operated on the island, more than one for each square mile! But these were small factories operated by mules and oxen. Later in the century windmills were introduced as the power source for grinding the harvested cane. Today dozens of brick and stone remains of these windmills dot the island's rural landscape. Only during the middle of the nineteenth century were these smaller mills replaced by large, modern, steam-driven factories. Now, even the large factories are closed (the last having shut during the early 1970s) and their rusting remains also can been seen throughout the rural areas of the island, although they were never so numerous as the smaller wind-driven variety.

ENGLISH HARBOUR

During the late seventeenth and well into eighteenth century, Britain battled especially the French and, to a lesser extent, the Dutch and the Danes for control of the eastern Caribbean. Antigua played an important role in the maritime military history of this period because of the naval facilities it developed around well-sheltered *English Harbour* on its southern coast. It provided the British with a secure base from which to initiate battles. A dockyard was built here to careen, overhaul, and provision British ships and to provide a safe haven during the hurricane season. It was an ideal port in which to keep the fleet in troubled times, since its enclosed harbor at *Freeman's Bay* is almost invisible from the open sea. Its narrow entrance was protected by *Fort Berkeley,* which was built on a narrow spit of land between 1704 and 1755. In addition, during sieges a strong chain-and-timber boom was drawn across the narrow entrance to English Harbour from the fort to the battery on the other side. The fact that Antigua was never relinquished by the British after this fort was built is a testament to its strength and design.

During the 1700s the dockyard was expanded to include more wharves, powder magazines, cisterns, and storehouses and it became the primary center for refitting the Caribbean component of the British fleet, which earlier had been provided for by the former northern colonies that are now in the United States. Its most famous resident (1784 to 1787) was British Admiral Horatio Nelson, who later commanded the British navy during the Battle of Trafalgar when the combined French and Spanish fleet was defeated in 1805. Because of his prominence, the English Harbour facility was named *Nelson's Dockyard.* In 1985 it was designated as a national park and has become Antigua's most important tourist attraction.

ST. JOHN'S

While English Harbour became the main military port on Antigua, the settlement of *St. John's* became its most important commercial trading center. Although this was not the first settlement on Antigua (the town of Parham was) it rapidly developed into the island's primary city by virtue of its well-protected harbor. Today it has a population of about 40,000 (representing almost half of Antigua's 85,000 residents). By the dawn of the eighteenth century, St. John's was a bustling frontier town of lumber stores, hardware shops, ships' chandlers, general stores, warehouses, grog shops, cheap inns, and houses of prostitution. Its streets were unpaved and became quagmires during the rainy season. Several outbreaks of yellow fever and malaria, plus a number of fires, earthquakes, and hurricanes added to the uncertainty of life during the 1700s and 1800s. *Fort James* was built in 1675 on a hill at the northern entrance to the harbor to provide protection.

Although most of the very old buildings in St. John's have been destroyed by natural disasters, a few remaining are worth visiting, in addition to Fort James. The town's *Court House,* on one of the corners of Market and Long streets, was built where an open-air market used to be in 1747. This stone building was also used for charity bazaars, official dinners, and other social events. Today, it is the home of the Museum of Antigua and Barbuda. *Government House* was built in 1788 at the corner of Newgate and Market streets in what was then the suburbs of the city. It is the home of the governor general of the island, who is appointed by the queen of England as a ceremonial administrator. *St. John's Cathedral* (its proper name is The Anglican Cathedral of St. John the Divine) was first constructed in 1683 but was rebuilt in 1722 and built again, after an earthquake, in 1845. It is at the junction of Church and Temple streets. This stone edifice is the largest building in St. John's and its twin towers can be seen from miles around. The *Public Market,* at the southern end of Market Street, is also worth seeing. It is especially busy on Friday and Saturday mornings and reminds one of the hustle and bustle that used to typify the city during the 1700s and 1800s. Here is where most of the fruits, vegetables, and meat produced on the island are sold, as well as some that is imported from nearby islands such as Anguilla, St. Kitts, and Nevis.

During the 1980s the Antiguan government invested millions of dollars dredging, extending, and building its cruise-ship port facilities located at the foot of St. Mary's Street in St. John's. In addition, in 1988 a new shopping and entertainment complex, *Heritage Quay,* was opened. Its facilities include duty-free shopping, restaurants, an amphitheater, and a casino. Somewhat less pretentious and much more picturesque, the shopping center named *Redcliffe Quay* was also opened during the 1980s as a tourist site. It is just south of *Heritage Quay* and also includes shops and restaurants. Redcliffe Quay is built around a number of St. John's older buildings that originally served as a slave compound, then called a *barracoon.* When the slaves were emancipated these buildings became warehouses for merchants until they were most recently renovated for tourism.

Field Trip around the Island, two and a half to three hours

The island of Antigua encompasses 105 square miles (273 square kilometers) of hilly and mostly brush-covered land. Its southwestern core is of volcanic origin, and most of the rest of the island is comprised of coral and limestone formations uplifted from the surrounding sea. Its highest point is *Boggy Peak* (1,319 feet or 402 meters). Although there is no longer any volcanic activity, the island is occasionally wracked by earthquakes. The last serious one occurred in 1974 (registering about seven on the Richter Scale), but numerous small shocks have been felt since then. A number of buildings were destroyed or damaged at that time, including the cathedral, whose cracks have only recently been repaired.

Antigua lies along the major track of tropical storms that traverse the Caribbean. Its wooden houses with corrugated metal roofs are particularly vulnerable to these storms. Ironically, however, the island's main chronic problem is dryness. An average annual precipitation of about 45 inches (113 centimeters) is simply not enough for a tropical location with a high evaporation rate. As a consequence, Antigua obtains most of its water supply from a number of desalination plants located along its coasts. Slightly more rain falls in the higher hills of the island's southwestern quarter. In terms of geology and climate Antigua resembles the Virgin Islands and St. Martin.

GREATER ST. JOHN'S, FIFTEEN MINUTES

Many of the roads on Antigua (except a few of the main ones and those in the city of St. John's) are not named, or at least their names are not posted on either signs or maps.

Leave *Heritage Quay,* drive east on St. Mary's Street for three blocks, turn left (north) on Market Street, and turn right (east) on Long Street. On the northwest corner of Market and Long streets you will find the stone building that used to be the old *Court House* and is now the Museum of Antigua and Barbuda. As you proceed east you will cross Temple Street. Look to your left (north) at the end of Temple Street and you will see *St. John's Cathedral* and its surrounding cemetery. Long Street will join Factory Road about two blocks east of the cathedral and you should continue eastward on it. At the juncture of these two streets you will see St. John's *War Memorial*, commemorating the First and Second World Wars and the Antiguans who fought in them. On the left is the *Antigua Recreation Grounds,* the island's main sports facility. One block east of this sports complex, on the left (north) side, is a fire station, and immediately behind it to the north is Antigua's main prison. Now you are in the suburbs of St. John's. About 0.25 mile (0.4 kilometer) straight ahead on the road and to your right (south) on a slight hill is the residence of the island's prime minister. Continue eastward on Factory Road and it will join with Queen Elizabeth Highway. Now you are in the exurbs of St. John's. Continue driving eastward for less than a mile (less than a kilometer and a half) past the small exurbs called Sutherland's on your left and then St. Johnston on your left and Potters Village on the right.

CENTER OF THE ISLAND, HALF AN HOUR

Once you have passed these settlements you are now in the countryside. Continue driving for about another mile (1.6 kilometers) and you will see the remains of a large sugar mill (the *Antigua Sugar Factory*), which was closed during the early 1970s. You can drive in and look at its rusting facilities. This mill was opened in the 1800s, during the height of sugar activity on the island. Many other unused facilities like this one are scattered throughout the island.

At the sugar mill a southward-running road intersects Queen Elizabeth Highway. Follow this road to the south. Less than half a mile (0.8 kilometer) down this road you find a set of large antennas and a telecommunications receiving dish on your left. This is the *Caribbean Relay Station*. It is where television programs from Great Britain and Europe are received and then beamed to other islands in the Caribbean, as well as being seen on television sets on Antigua. Just to the south of the relay station is a new residential area named *Light Foot*. This is one of the more affluent neighborhoods on the island. Most homes here are built on one-acre (two-and-a-half-hectare) lots and the area is less than 5 miles (8 kilometers) from downtown St. John's. About a mile (1.6 kilometers) farther south is the village of *Sea View Farm*. It is noted primarily as a residential area for people who work in St. John's and is famous for its clay pots made by local women. These pots are used for storage, flower pots, and *coal pots* (clay pots that function like an oven).

Continue southward to the village of *Liberta*. It got its name because it was built on land given to freed slaves after their emancipation between 1834 and 1838. Other villages on Antigua originated in similar fashion, for instance, Freeman's Village southeast of where you traveled through Sea View Farm. As you enter Liberta from the north you pass under an arch constructed of steel rods. It was built in remembrance of the island's independence in 1981. All of the land around Liberta, as well as most of the land along the roads on which you have traveled to this point beyond St. John's, was covered by sugarcane until about thirty years ago. Now some of the land is used for vegetable and fruit production, but most is either unused or is being grazed by goats or a few cattle and sheep. If you look carefully you can see the scars on the landscape, where the natural vegetation has been cut over.

ENGLISH HARBOUR AND ENVIRONS, ONE HOUR

Continue southward to the settlement of Falmouth on a small hill above *Falmouth Harbour.* From here you can get a good view of the harbor below toward the south. This has become one of Antigua's prime yachting meccas. During the last week in April and the first week in May each year *Antiguan Sailing Week* is held here. It is an international event that has turned into one of the top ten sailing regattas in the world. Thousands of spectators travel here to watch the races and join the parties afterward.

As you travel south and east of Falmouth Harbour you will come to *English Harbour* and *Nelson's Dockyards.* The dockyards and the protective land at higher elevations above containing the fortifications of *Shirley Heights* and *Blockhouse Hill* and also *Clarence House* are must visits for anyone traveling to Antigua.

THE SOUTHWESTERN ISLAND, HALF AN HOUR

Retrace your path to English Harbour back northward through Liberta to the road that runs toward the west through the village of Swetes. Between the villages of John Hughes and Old Road is *Fig Tree Hill Drive.* This is a popular drive because it is the wettest place on the island (receiving about 60 inches, or 150 centimeters of rainfall annually) and, as a consequence, the vegetation here is more luxuriant than elsewhere. It is often described as a rain forest, but technically this is incorrect because it is not at all a natural landscape. The trees here are almost all fruit trees that have been planted. The rainfall is seasonal, so there is a pronounced dry season, and it is not nearly enough to allow growth of the kinds of trees that typically grow in a rain forest (like in the Amazon Basin and along the Caribbean coast of Central America). The kinds of trees here include such varieties as the mango, orange, guava, coconut palm, breadfruit, papaya, passion fruit, and banana. In fact, when Antiguans use the term fig they are referring to what most of the rest of the world calls banana. Thus, Fig Tree Hill Drive would probably be called Banana Hill Drive elsewhere in the tropics.

To the north of the road you are traveling on are the highest elevations in Antigua (800 to 1,300 feet, or 244 to 397 meters, above sea level). These are the volcanic hills and the highest hill is Boggy Peak (it is the one with a radio antenna on top), about a mile and a half (2.5 kilometers) north of the village of Urlings.

As you travel westward on the road from the village of Old Road you will view some beautiful beaches and you can also see toward the south the island of Montserrat and, on a clear day, St. Kitts. There are also several small pineapple farms in this area. As the road bends to the north you will be heading back toward St. John's. Along the way and to your left (west) are some of the largest resort complexes in Antigua, including Jolly Beach Hotel, Hawk's Bill Hotel, and the Royal Antiguan. Because the road curves inland here you will not be able to see these very well unless you take the time to travel the short distance to them along one of the side roads. Along the way, as you pass through the town of *Ebenezer*, you will see a number of small houses that all look about the same on the left behind an elementary school. This is Antigua's largest low-income housing project. It was built with government funds for poor people who used to live in substandard housing. The people living in these houses buy them at below-market prices and use low-interest long-term mortgages provided or arranged by the government.

BACK TO ST. JOHN'S AND THE RESORT AREA TO THE NORTH, HALF AN HOUR

Continue northward through the eastern edge of the city of St. John's and follow the coastal road north to see the resort developments along the coast just north of the city. This is the largest single concentration of resorts on the island. The beaches here are beautiful and the area is close to both St. John's and the airport. The resorts built have entertainment centers, sometimes including casinos. Follow this road northeastward to Cedar Grove and then turn south and follow the road back to St. John's. On the way southward you will see the *Antigua Police Training School* on your right (west). This is where all Antiguan policemen are trained. Just to the south you will pass through the Agriculture Department's main experiment station. And farther along the way south, to your right (west) you will pass by *The West Indies Oil Refinery,* which is now closed. Just before you get to this refinery, you will see a very modest little house back in the brush to your right (west). This is the house that Antigua's prime minister, Vere Cornwall Bird, was raised in. He got his start in politics when he helped create Antigua's first labor union in the 1930s. These activities subsequently produced the Antigua Labor Party, which Mr. Bird has headed ever since.

Continue on your return to St. John's and take High Street back to Heritage Quay, where the cruise ships dock.

Barbados

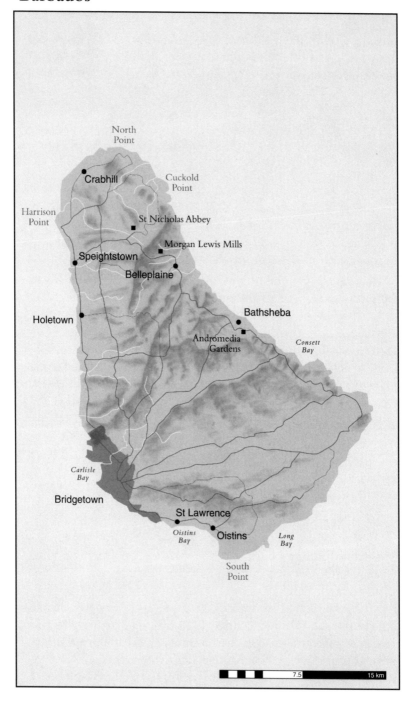

△ Day Five

BARBADOS—WHITHER BIMSHIRE?

Barbados, a mere 166 square miles (432 square kilometers) in area, is the easternmost of the Antilles, separated from the Windward Island arc by the Tobago Trench. The majority of the island is composed of Pleistocene coral limestone, overlying a highly folded Tertiary series of sedimentary strata which only surfaces in the Scotland district in the northeast of the island. A distinctive cuesta, *Hackleton's Cliff,* marks the rim of the Scotland district. The remainder of the coral landscape displays subdued karst topography, with dry valleys and dolines etching the scenery and with a series of uplifted coral cliffs paralleling the western coastline. The original primary vegetation of tropical semi-evergreen and deciduous forest was almost completely removed to make way for the growing of the island's "brown gold," sugarcane. Indeed, the Portuguese gave the island its name, Los Barbados, after the appearance of the banyan, or bearded fig trees that lined the Caribbean coast of this island. Apparently, the island was not inhabited by Arawaks or Caribs when discovered, nor claimed by Portugal, and it remained for the British to settle and develop this island from 1627 to 1966, as one of their most profitable sugarcane plantation economies in the region.

First, pioneer farmers and estate owners with indentured servants from rural Britain attempted to grow and sell tobacco, cotton, and ginger. Then with Dutch assistance, by way of Brazil, sugarcane was introduced as a plantation staple product to Barbados. Twenty years later the island was fully developed into small sugar estates, more than 700 at their zenith. Many pioneer settlers left—some for the U.S. mainland, especially Virginia and South Carolina; others returned home; others moved elsewhere in the Caribbean; and a residual remained to form a low-class white strata of plantation overseers, merchants, and domestic servants. The introduction of slavery, to satisfy the enormous labor demands of these highly profitable sugar plantations, completely changed the demographic profile of the seventeenth-century population of Barbados. In 1640, the population consisted of approximately 37,000 white Europeans and 6,000 African slaves. By the year 1786, there were only 16,000 whites, while the African population had increased to 62,000. Continued increases in the African population made Barbados one of the most densely populated countries in the world, so that by 1921, African descendants numbered 180,000, while the "white" and "colored" minority remained more or less fixed at 15,000. Today, in the 1990s, this 14- by 21-mile (23- by 34-kilometer) island has a resident population numbering approximately 260,000; ninety-two percent are African-Caribbean, four percent white European, and four percent are of mixed race or minority ethnic origins—East Indian or Levantine, for example.

Under slavery, the African majority was forcibly detained. Escape, an exceptionally high-risk option, did at least have an abiding "demonstration effect" and most certainly initiated the tradition to "flee the plantation" once emancipation was a reality. The birth of an emigration tradition for Barbadian rural underclasses, therefore, can be attributed to this plantation/slavery experience. After emancipation, moving to an undeveloped interior was not possible in this overcrowded and ecologically stressed small island. For the adventurous or desperate, movement away from the plantation necessitated an international move off the island. The majority, however remained on Barbados, forced to rely on laboring opportunities on the plantations, even after the aborted apprenticeship scheme and their acquisition of "freedom."

Soon economic hardship, ecological deterioration of land resource bases, environmental calamities, and the restructuring of the international mercantile and industrial economic system in the late nineteenth century combined to render the Caribbean susceptible to penetration and exploitation by North American capitalist interests. Corporation agents actively recruited labor in Barbados, and its colonial government remained relatively passive in its acceptance of the situation. Offers of a paid-passage and wage enticements were sufficient to encourage the more adventurous, or the more needy and more desparate, to undertake temporary labor contracts overseas in Panama, the Hispanic Caribbean, or farther afield in South America. Hence emerged an international circulation tradition among Bajans (as these nationals are called), which changed the geographical scale of their opportunity field(s) to encompass extra-regional destinations. The persistence of destitution at home, together with the demonstration effect of successful returnees and the social pressure on those who stayed behind to also demonstrate success, prompted rapid conversion of this selective strategy to a mass phenomenon.

Later generations of the Barbadian underclasses were to receive continuing reinforcement of this circulation strategy, as the need for such a pliable and temporary labor force was proved to be advantageous to capitalist expansion, both in the metropolitan mother countries and most recently in North American agricultural and metropolitan economies. The local social and economic environments perpetuated the process, also. Folk-tradition reinforcement aided and abetted continuance of the habit of viewing the outside world as the place of opportunity and advantage. The openness of these insular societies to metropolitan influences, in part reinforced by the cross-currents of movement of Caribbean people, contributed to its maintenance. Barbadians went far and wide in search of overseas opportunities—to South America, to Central America, to Europe, to North America, as well as within the Caribbean. These emigration and circulation traditions have long been a "safety-valve" suppressing population growth rates, while extra-regional subpopulations have emerged as extensions of this island community.

Substantial overseas populations of Bajans continue to consider themselves as Barbadian, and these overseas enclave communities—in such cities as New York, Toronto, and London—retain strong ties with their Caribbean home as reflected in their visitation patterns to this Caribbean island.

For 300 years, Barbados was a sugar island par excellence. As late as 1962, for example, more than ninety percent of the island's exports were of raw sugar, rum, and molasses. At its zenith, more than seventy percent of the island's land surface was cultivated in cane, with another fifteen percent under other agricultural uses: pasture and vegetable-growing. Since gaining political independence in 1966, a major goal of successive Barbados Labor Party and Democratic Labor Party governments and their various development boards has been to diversify the island's skewed economic structure and dependence on sugarcane exports. Sugar, rum, and molasses remain the major agricultural export commodities, in large part due to the preferential quota systems of Lomé trade agreements with the European Economic Community. Commonwealth trade agreements with Canada and the bi-lateral Caribbean Basin Initiative arrangements with the United States, crop diversification, renovation of an artisan fishing industry, light industry, export-oriented diversification, and development of the tourism and producer services industries—each have contributed to an expansion of Barbados's economy that makes this small island one of the most diversified of Caribbean economies.

Tourism gradually replaced sugar as the island's main economic sector. Agricultural diversification has not been without its problems, however. Experiments to develop Sea Island cotton as a valuable staple failed. Over the years since political independence, agricultural-development efforts to introduce marketable crops among the small farming communities have had a checkered history. Onion growing failed, for example. Gradually, however, the annual bill for imported food has not increased so dramatically as local market structures have improved delivery services and more and more managed to satisfy the consumption patterns of tourists and locals alike. Dairying and raising fruit trees—paw-paws, avocados, and citrus—have grown to partially meet domestic demand. The establishment of a hierarchy of supermarket chains in and around the urbanized area of Bridgetown and its south- and west-coast tourism corridors has helped expand the domestic market for green vegetables, tomatoes, and yams and sweet potatoes. The proliferation of restaurants to serve tourists and the expanding

Bajan middle class has also contributed to service efficiencies in the agricultural sector. The import bill for food is still substantial, but the island's high degree of dependency upon foreign food imports has been at least reduced from its high pre-independence totals.

The two half-day itineraries are designed to introduce geographers to the island's physical and environmental landscapes and to the contemporary influences which have wrought such dramatic changes since this Caribbean "little England" (also colloquially known as Bimshire) gained political independence in 1966.

Bridgetown to St. Nicholas Abbey and Bathsheba, a half day

Barbados is well sign-posted, unlike other Caribbean islands. This road trip takes us on a northern route out of Bridgetown, through the leeward parishes of St. Thomas, St. James, and St. Peter; across to St. Nicholas Abbey in St. Peter, then south along the east-coast road through Belleplaine and Bathsheba in St. Andrew and St. Joseph, respectively; returning through the island's interior, the parish of St. George, to St. Michael and Bridgetown.

DEEP-WATER HARBOR TO THE UNIVERSITY OF THE WEST INDIES, TWENTY MINUTES

We leave the Deep-Water Harbor entrance in Bridgetown and turn left at the first turn toward Fontabelle. Now we enter Barbados Industrial Development Corporation, *Pelican Estate,* the island's earliest enclave industrial estate, which includes Caribbean Data Services (AA data processing), Confederation Client Services, and tobacco and garment-assembly factories. To the left we can see the large bulk-sugar warehouse. Beyond this is a newer Industrial Development Corporation extension, the Harbor Industrial Estate, with its mix of export-oriented plants. Turn right at the roundabout and circle the *Deep-Water Harbor* and the small-boat, inner harbor where two Jolly-Roger tourist ships mingle with inter-island transports. We soon reach a set of traffic lights, where our route takes us left onto the Spring Garden Highway and out of Bridgetown.

Along Spring Garden Highway are well-preserved chattel houses on the right. On the left is a large grain-storage elevator. On the left we come upon a typical Caribbean beach, with calm waters and plenty of local swimmers. Farther along the highway where there is open space and grass, pick-up games of soccer are common early morning activities. This is a sport-mad island! Cricket is the national sport, though summer is the season for soccer and tennis, as well as cricket—all sports having competitive leagues. Continuing west along the highway, on the left we pass a major oil-storage facility operated by Shell Antilles. The Spring Garden Highway climbs up a short incline and we arrive at a roundabout, where we turn right off this west-coast road toward Eagle Hall, climb a much steeper hill, and reach the *University of the West Indies,* Cave Hill campus, which stretches out on the right.

Continue on the Eagle Hall Road through to the University campus, past dormitories, university buildings, and sports fields on the left, an upper middle-class residential area and several research institute buildings on the right, and arrive at the Eagle Hall traffic lights. Crossing these we find ourselves on the new George Cummings Highway and go over the brow of the hill to the first roundabout, the "Simpsons Suzuki" roundabout, where we turn left on Highway 2A to go north. There are plans here for an industrial estate to be developed and the Suzuki assembly plant is evidence of this. In addition, the long-abandoned cane fields on either side of the Cummings Highway, the "C" in the "ABC" highway, are currently being converted to residential subdivisions and there are signs of building affluent housing in the fields to the north of the roundabout.

ST. MICHAEL TO ST. PETER, TWENTY MINUTES

Turn north on Highway 2A toward Clermont. From this point on, the road is a two-lane highway, quite narrow with no shoulders. The affluent housing along this road is a response to the increased accessibility to all parts of Greater Bridgetown afforded by the ABC Highway. Some of the cane fields on either side of Highway

2A are still used for agriculture; others are left uncultivated with the expectation that they will be sold and converted to residential use. *Warren Heights* is typical of this recent suburban development. The older residential housing along this road is interspersed with newer building. Some of the old plantation houses are also undergoing renovation and occupation, only a few still accommodating the planter family or plantation overseer. Housing stock varies from wooden chattel housing, to modest concrete-block housing, to affluent two-story modern detached residences, a few of the latter adorned with their most recent status symbol, a huge satellite dish. At Bagatelle Terraces, the plantation house has been converted to a private business. As we travel north on 2A, some fields are planted with yams and sweet potatoes, but also some cane. We parallel one of the several cliff-like ridges, where groves of mahogany trees occupy ravines, or remain as clumps of shade or cover alongside the road. The uplifted shoreline and steep cliff to the right of the road are tree-covered, with observation towers at the crest. Here, there is no sign of residential development. To the left and closer to the sea, the ridge is built-up with housing. All around Bridgetown, these cliff sites have become prime residential property.

Highway 2A passes through *Arch Hall,* one of Bimshire's "freedom villages," and then meets Highway 1A at a crossroads, this road going down to Holetown on the west coast. We proceed north into St. James and the new Port Vale factory, built in 1981. There is a small sugar museum at the *Port Vale* factory. Mahogany stands line the side of this old road. Here, there is little residential development, though some evidence of agricultural diversification. Beyond Port Vale, the road dips into a hairpin bend and crosses Lancaster Gully, a mahogany-filled ravine, a collapsed limestone cavern similar in formation to karst dolines. Eventually, this old narrow highway opens out to the newly widened and improved Highway 2A. Occasionally, in the open fields at the side of the road, coral–limestone sinkholes (locally known as "sinks") can be seen. This new highway bypasses Westmorland Village. We continue north. Going toward Bakers, settlement becomes more evident, as we come within the outer reaches of Speightstown. A few

National Conservation Counsel (NCC) gardens can be seen in the communities. On the right are more mahogany stands since the highway is now running at the base of the inland cliff. We reach a turn to Mullins Bay at Bakers, and continue onward.

Entering *St. Peter,* we are presented with a wide panoramic view of Speightstown to the left. There's another sugar factory in the middle distance, and far off are the huge towers of the *Arawak Cement Works,* a joint industrial venture between the Trinidad and Barbados governments. A reconstructed windmill, or millwall as the locals call them, is on the left coming toward Mount Brevitor and Mile and a Quarter.

MILE AND A QUARTER TO MORGAN LEWIS, ONE HOUR

Soon, at Mile and a Quarter, we join the original Highway 1 coming out of Speightstown and turn right to follow it inland. Going up the hill to Mount Brevitor, we pass through our first cut through these inland, limestone cliffs, invariably circling around a plantation house on the bluff above. We continue along old Highway 1 and eventually meet the new Highway 1, the Charles Duncan O'Neil Highway, where we turn right again. We pass Pleasant Hall House on the left. Now with cane fields on either side, we are in the interior of the island. Arriving at the turnoff of old Highway 1, which continues on to Farley Hill and on to Belleplaine, the newly constructed highway "goes straight" (veers left) to the Abbey via Diamond Corner. At Diamond Corner, a tidy little nuclear village, we turn right. Breadfruit trees and croton-decorated gardens enhance this village's appearance, and many of its modest houses appear carefully maintained.

Quite soon, we enter a grove of mahogany trees, and in its midst turn left into the grounds of *Nicholas Abbey,* a Jacobean great house. Admission is $5.00 Bds. Here one can view a film recounting earlier times in Barbados: sights of early Bridgetown; the Careenage; the hustle and bustle of carts, people, sugar casks, and produce being shipped out of the island's main port; sights of the sugar plantations and factories as they were when King Sugar reigned supreme as the main livelihood of all Bajans. A stop here lasts three-quarters of an hour. There's a modern windmill on the

St. Nicholas Abbey, Jacobean Great House, Barbados. Photograph by Dennis Conway.

left as we leave. Leaving Nicholas Abbey, one of the few remaining three-story plantation houses, we turn left and continue up through the mahogany grove to suddenly emerge in the open at the top of Cherry Tree Hill.

This is the Boscobel Road to Morgan Lewis and Belleplaine. Stop at the crest top and admire the spectacular view—there stretching to the southeast is the windswept Atlantic coast; nearest is Walkers Savanna; beyond is Chalky Mount and other parts of the Scotland district; farther on are the perched rocks of Bathsheba and all points south to Consett Bay; and, on a clear day, Ragged Point in the far distance. Our path winds down the hill, and more and more vistas open up on the descent. At the roadside are wired coral-stone blocks, a conservation service attempt to reduce

Morgan Lewis Mill, Barbados. Photograph by Dennis Conway.

run-off erosion. The sides of the steep hill slopes are terraced here, and this area is one of the early conservation districts of the island, where considerable attempts have been made over the years to reduce erosion and stabilize the hillsides, to prevent landslips, landslides, and road deterioration. On the left as we go down farther, signs of conservation efforts are everywhere. We drop down to Morgan Lewis, where a feed-lot is being operated. On the right is the *Morgan Lewis Mill,* a Barbados National Trust–managed site ($2.00 Bds. admission fee).

WALKERS SAVANNA TO ANDROMEDA GARDEN, HALF AN HOUR

We continue down the hill and on toward Belleplaine, passing through Greenland, where we join Highway 2. On the left are the sizeable dunes of Walkers Savanna, long a source of sand and building materials. Now, conservation efforts have reduced the amount of unregulated quarrying of these inland sand dunes. We pass by St. Andrew's Church on the right. Here is a flat plain given over to grazing of staked goats and sheep. At the church in *Belleplaine* we turn left, pass a banana plantation, and enter this regional center. At the town's main supermarket, we turn left onto the eastern-coast road and pass more sports fields, including a basketball court—the first visible evidence of this newest Bajan sport craze which threatens the monopoly of cricket as the country's national pastime.

Reaching the windblown coast, there are salt marshes to the left and the jagged and wind-etched ridges of Chalky Mount towering above on the right. Sea grape bushes there fix the sand dunes and provide some wind protection along this coastal road. The open trade-wind exposure provides a lasting impression of wildness and salt spray, best experienced by walking along the sand. South on this coast road, there are a few beach houses. We continue through Barclays Park on to *Cattlewash*—yes, their cattle are used to being washed in the sea—and stop here to take in the salt air of this truly windward coast. Trees are shaped by the wind, the rocks of the bluffs above are etched.

At the southern end of Cattlewash beach are family vacation homes, most owned by old Bajans who have been coming to this

Bathsheba and northwards, Windward Coast of Barbados. Photograph by Dennis Conway.

side of the island for generations. We pass the Kingsley Club, a white Bajan preserve, on our way out of Cattlewash and up a very steep hill. Turn left across the same old bridge; then at the top of the next hill, turn left and immediately left again to drop down into *Bathsheba,* the next village down the coast. Houses here are still wooden, in the main, though sporting paint. We turn left down Reliance Road toward the coast and wind our way into the bay area. The lovely vista to the left reveals houses perched on the hillside and some casuarina trees shading the houses. Pass the Wesleyan Holiness Church and traditional old wooden houses and arrive at the foreshore, where the pedestal rocks protrude out from the shore. The NCC garden between the road and the seaside is part of a successful beautification campaign. On the southern edge of the town, we again swing around a hairpin bend, past a brightly painted and mural-decorated community center, up the hill out of the village. Halfway up the hill out of Bathsheba we come upon *Andromeda Gardens* on the left, a National Trust property. A tour (thirty minutes, $5.00 Bds) around Andromeda gives you the chance to examine the varied tropical trees and imported plant species that characterize the southern Caribbean.

HORSE HILL TO BRIDGETOWN, TWENTY-FIVE MINUTES

Leaving Andromeda Gardens go left out of the grounds and up the hill about 100 yards (100 meters) to join Highway 3. Turn right onto the main road. From here, Highway 3 climbs inland toward Hackletons Cliff. Climb a long, steep road to the top. Now we come to a road-improvement project in *Horse Hill* village: Gullies have been modified, and the village cricket field converted to a basketball court. At the crest of the hill is a massive cut; the exposed limestone rock is falling away on the steep sides, and crumbling—the surface needs to be treated. After the crest we drop down on the cuesta of the escarpment, through cane fields. On the left we pass *Blackman's plantation house* with its pink walls, still inhabited by its original planter family. On either side there is cane in various stages of planting—just as it has always been. We go down the first inland cliff, round the usual hairpin bend with the blackened coral-stone cliffs (some decorated with motifs) and pass another small garden on the right, flamboyant trees above; we continue and drop down past Andrews cane factory. On the left is another example of a restored and gentrified plantation house. An elegant avenue of Royal Palms lines the drive of another great house on the right just past Andrews. From this point onward, we begin to see the development of high-quality residential nuclei on cliff bluffs as we come within the Bridgetown commuter shed in this interior part of St. George, nearing Fisher Pond and Sweet Vale.

Just before Market Hill our route comes to a multiple junction with Highway 3B; we turn left on 3B toward *Grove Agricultural Station*, and at Grove turn right and south to join Highway 2B toward Gun Hill. At Grove, sugarcane species are bred. This is a cut-off road joining Highway 2B turning right. In Newbury village we turn off 2B, then turn left again following the signposts to *Gun Hill.* The first building at Gun Hill is a recuperation station in poor repair, then a Signal station. The gardens are being tended by NCC; the station is managed by the National Trust. We stop here to take in the view of St. George to the south, St. Michael and Greater Bridgetown to the west, and farther to the northwest the series of inland cliffs up the west coast in St. Thomas.

We leave via the road that curves left at the bottom of Gun Hill with its lion on the right, the site cleared by the NCC. Back on Highway 2B, we continue south turning right. Almost immediately we drop through another cut in the limestone cliff to a straggling open village of *Glebe,* with its mixed housing stock and a little estate nestling to the left. Developments here demonstrate the community-level infrastructure. The road passes into a village square, with St. George's Church on the left, a cricket field in the center. Large mahogany trees line the left-hand side of the road. At the crossroads, we join Highway 4.

Passing the Roberts manufacturing building, a modern feed plant, at the stop sign we take a left turn and join the ABC highway near the old Belle factory on the far right side of the highway. Turning left onto the ABC highway we go up to the Emancipation statue and from there, turning right and onto Two Mile Hill, go down Government Hill into Bridgetown. On Two Mile Hill we pass a government project under construction. Next to it is the official prime minister's residence at Illaro Court on the left-hand side, and the headquarters of the Caribbean Broadcasting Union on the right, occupying the former residence of King Ja-Ja. There is a yellow mayflower tree on the right. On the left is an office building of Barbados Community College and the BARTEL Headquarters on the right. Still to the right are middle and lower class residences of the Ivy and Government Hill communities. On the left is a secondary school, the Florence Springer Memorial, with a railed pedestrian walkway. This walkway lies on the outside of the spacious grounds of the governor general's residence, Pilgrim House, currently occupied by Dame Nita Barrow. Forced to turn left following the traffic flow and the traffic lights, we bypass the imposing entrance to the governor general's residence and turn right through the roundabout onto Belmont Road, with Belleville, an old high-class Jewish neighborhood, on the left; then continue toward Queens Park, the former residence of the major general of the Royal Forces of Barbados. On the left are the new Ministry of Education buildings nearing completion on the old Queens College site. We pass Fort Royal Garage, the former site of the St. Michael's Almshouse, Church Village, on the right. Next is the historic St. Michael's Cathedral, built in 1665. We soon come to Trafalgar Square, Bridgetown's ceremonial site with its historic statue of Lord Nelson. Trafalgar Square was enlarged to its present size in 1807. The one-way traffic system circles us around Trafalgar Square and along Wharf Road and the Careenage through the business heart of Bridgetown to complete this tour.

Bridgetown along the South-Coast Tourism Zone

Starting from Pelican Village on the Princess Alice Highway, we pass the newly constructed Bridgetown Fisheries complex. Passing through an old warehouse district, our one-way system turns left onto Prince Alfred Street and passes the ornate Mutual Life Building on the left, now occupied by Barclays Bank. We turn right down Broad Street, the town's main commercial artery. Passing through Trafalgar Square, the one-way system takes us past the Department of Inland Revenue building. We turn right and cross the bridge across the Careenage, passing the Fairchild Street bus terminal, now modernized, on the right. Still in a one-way system, pass the Empire Cinema on the right hand and on the left a minibus station on Probyn Street. This area is also known as Golden Square, the scene of the early 1937 disturbances. Now on Bay Street, there are a number of quaint buildings with overhanging verandas. Important buildings to be seen on Bay Street are: The Harbor Police Station, Martineau House, the Child Care Board, the Old Adjutant Quarters, the Roman Catholic Cathedral, and Clarence House, now occupied by the Ministry of Education. Soon there are the first signs of restaurants and establishments catering to tourists. The entire stretch of Bay Street displays a fine array of

Middle-class housing along South Coast Road, Barbados. Photograph by Dennis Conway.

old vernacular housing, intermingled with gentrified tourist businesses. This area experienced its change from agricultural use to residential development as early as 1805.

Where *Government House* and the *Esplanade* are on the left, Carlisle Bay is on the right. Further along on the right is the coral-walled Yacht Club, until 1966 exclusively for "whites only." Then we pass the turn off to Needhams Point, site of one of Barbados's first commitments to international tourism, the Hilton Hotel, and up a gentle incline to the *Garrison Savannah* and the barracks of the Barbados National Regiment. The Garrison is a hub of sports activities: horse racing, cricket, rugby, soccer, even kite flying. The old elite housing around the Garrison has been converted to small business use, and the barracks now house the city-planning department. *Hastings* police station occupies the

Flying Fish Fleet in Careenage Harbor, Bridgetown, Barbados. Photograph by Dennis Conway.

right-hand corner of the Garrison as we continue along the coast road. This heralds the beginning of the southern coast zone, where for the next several miles a poorly regulated mass-tourism landscape has come to dominate the original residential and commercial mix. Some of these picturesque old houses have been renovated, and some of the character of original balconied wooden- and coral-stone housing has been retained. But all too often, newer concrete-block apartment buildings, fast-food restaurants, bars, small boutiques, banks, and tourist-shopping malls crowd together along this coastal road, creating a lasting impression of chaos.

From Hastings, past Rockley, through Worthing and on to Dover, the names reflect "little England." The scenery is in part robbed of its Caribbean flavor by the density of tourist apartments, hotels, cafes, and restaurants. Some of the earliest Barbados hotels, like the Ocean View, still compete with their newer opponents, where video arcades and Kentucky Fried Chicken franchises are found to cater to the mass tourist from Europe, Canada, the United States, as well as elsewhere in the Caribbean. On the left is Plantation supermarket, which used to be Goddard's supermarket at the bottom of Rendezvous Hill, an early retail node, now barely distinguishable in the almost continuous ribbon of development. Where we catch glimpses of Worthing Beach on the right, the *Graeme Hall Swamp* is to the left—a haven for migratory birds, and nesting site for white cattle egrets, or tick birds as they are commonly known. We turn off this coastal road down St. Lawrence Gap, to pass into Dover along the St. Lawrence Coast Road. Here is a tourist enclave, where tourists can feel quite separated from the local people. Eventually, and just before the convention center, we turn left at Dover Cricket Club and wind our way through chattel houses, back onto the coast road. As we emerge onto Highway 7, you may see a boat builder, still handcrafting his fishing launches on the side of the road. Turn right and come up to the roundabout where you turn right on Maxwell Coast or "Bottom" Road (to distinguish it from Maxwell Top Road). In five minutes of relatively unchanged residences, enter the fishing village of *Oistins*.

Coming out of the Oistins fishing market we turn right and soon come upon a cluster of government regional offices and a district hospital and administrative buildings on the right; this is one of the island's regional parish centers. Leave Oistins as Highway 7 climbs a couple of steep bends. The original road from Bridgetown to the airport, past Thornbury Hill and Durants, is on the left; on the right is more residential development, and stretching below is the extensive residential development of Enterprize and Atlantic Shores toward Silver Sands (referred to as Chancery Lane). This whole plateau has been built up in the last fifteen years. Nearing the airport, there is some agricultural diversification in the experimental station on the right. Then we come upon the airport industrial estate on the right, and turn around, passing through the departure lane of Grantley Adams International Airport.

Our return trip takes us along the ABC highway, first along the Tom Adams Highway, through Newton, Kingsland, and Warners. Then we go straight at Barrow roundabout to turn right past the tall BARTEL skyscraper, through the traffic lights and on over the brow of Upton Hill through Wildey, past the Samuel Prescod Polytechnic and the Caribbean Development Bank headquarters, in the Wildey estate, and on to the St. Barnabas Emancipation statue, Bussa roundabout and the Errol Barrow Highway. Here we retrace the route down Government Hill. This route, like all the other highways the ABC highway intersects on its east-to-west traverse of the outskirts of the city, takes us back to Bridgetown.

Guadaloupe

Anse-Bertrand

Port-Louis

Grand Cul-de-Sac Marin

Moule

Baie du Nord Quest

Grande Anse

Grande Terre

Saint-François

Baie-Mahault

Pointe-À-Pitre

Gosier

Petit Cul-de-Sac Marin

Rivière Salée

Vernou

Petit-Bourg

Basse Terre

Cascade de Escreviasses

Mt Soufrière +1463m

Bouillante

Anse à al Barque

Saint-Claude

Marigot

Gourbeyre

Trois-Rivières

Parc Archeologique des Roches Gravées

Basse-Terre

Fort Charles

Vieux Fort

Sainte Louis

Marie Galante

25 km

12.5

△ *Day Six*

LA FRANCITÉ DES PETITES ANTILLES: GUADELOUPE—LES VISAGES CONTRASTÉ

In an earlier phase of Caribbean regional history, French influence was strong, France warred with Britain over possession of many of the islands, and French settlement and culture became very much part of the Europeanization of the region. Today, few islands retain formal ties with France, but the French Creole cultural traits are widespread. Local dialects or patois in several islands are French–English–African mixtures; French architectural styles proliferate throughout the region; aspects of French cuisine remain in local culinary traditions; Roman Catholicism mixes with Anglican, Huguenots, and even Evangelical religious traditions among Europeanized Christian beliefs. Haiti's successful Jacobin rebellion of independence from French colonial rule scarcely purged French cultural influences from that Caribbean nation, and more recent Haitain diasporas throughout the region and farther abroad to North America and Europe have tended to replenish Caribbean French Creole traditions and cultural bases.

Two Caribbean islands remain, each politically and administratively Départements D'Outre Mer, effectively political divisions of mainland France. Martinique is one prefecteur, Guadeloupe the other, the latter having administrative responsibility for several small island dependencies: Saint Martin–Sint Maarten, an island shared with the Netherlands; St. Barthélemy, or St. Barts as it is commonly known; and the nearby miniscule islands of Marie

Galante, Desirades, and Les Saintes. Following the Algerian debacle, Charles de Gaulle's France shed itself of its vast colonial empire in Southeast Asia and North and West Africa, but retained sovereignty over four "strategic" colonial possessions: Reunion, French Guiana, Martinique, and Guadeloupe and her brood of Caribbean dependencies. These Départements D'Outre Mer remained as overseas provinces—the people elect deputies and prefects who serve in the French Parliament in Paris, they enjoy all the legal and constitutional rights of French citizens, and all the French social welfare programs and educational programs open to mainland citizens are ostensibly available to these Caribbean "citoyens." Distance and especially cost separate these overseas French from their mainland cousins, however.

Guadeloupe is physically two islands separated by a shallow inlet and an isthmus, with the Rivière Salée, a mangrove-filled swamp, constituting this narrow linking land-bridge. The westward island *Basse-Terre* is volcanic, with an area of 364 square miles (946 square kilometers). The highest of its rugged volcanic peaks is *Mount Soufrière*. At 4,800 feet (1,463 meters) elevation, it is also the highest peak in the Lesser Antillean arc. To the east is *Grande-Terre* (219 square miles, or 350 square kilometers), a low-lying island of undulating subdued relief formed from a coral limestone cap. The dramatic differences in geologic structures and relief are accentuated by climatic differences between Basse- and Grande-Terres. Low-lying Grande-Terre has an east-to-west gradient of climate zones, from humid, through semi-humid to semiarid subtropical regimes. On rugged Basse-Terre there is a rain-shadow effect on the western coast, but on the windward slopes and at higher altitudes precipitation totals are high, and tropical humid forest is, or was, the primary vegetation.

The two islands, each a "wing of the butterfly," also differ in their agricultural bases. Grande-Terre is predominantly a monocultural sugar-plantation landscape, where even the approximately twenty percent of subsistence farmers on Grande-Terre rely on their sugarcane crop for their livelihood. Here too, cane is grown for eighteen months, the grande culture, to enrich the sucrose content before harvesting. On Basse-Terre, subsistence farming

Point-à-Pitre, Guadeloupe, French Antilles. Photograph by Dennis Conway.

mixes with larger estates. The latter grow bananas, coffee, mangoes, and breadfruit; the former grow tree and garden food crops for subsistence as well as local markets. Guadeloupe's corporate agricultural sector has stayed with the production and export of bulk sugar to the mainland and within the European Economic Community, while Martinique chose to refocus its sugar industry to concentrate on exporting rum and molasses. In light of the prevailing downturn in world bulk sugar prices, Martinique's formula appears to have been more flexible than Guadeloupe's.

Point-à-Pitre, the commercial capital of Guadeloupe, has a population exceeding 100,000, whereas the administrative capital of the department, *Basse-Terre,* is relatively small, with only approximately 14,000 inhabitants. As the commercial heart of Guadeloupe, and strategically located in the Caribbean, Point-à-Pitre's port has been developed, enlarged, and modernized. Although such infrastructure developments have come later to Guadeloupe than to Martinique, its commercial future appears bright. French, European, and international financial investment has certainly altered the modernizing landscape of Guadeloupe. Tourism and hotel development is one focus, the transport and energy infrastructures have been modernized, and offshore industry invited. Point-à-Pitre, once a somewhat tardy and unexceptional port city, has transformed itself, in large part a result of successful political maneuvering by mayors and city councils to garner mainland resources.

Basse-Terre, one day

Leave Port Autonome de la Guadeloupe, where cruise liners dock on Quai 3. Travel northwest along Quai Foulon and then northward on Quai Lefebre to Boulevard Chanzy, the modern four-lane highway leading out of Point-à-Pitre. Turn left onto the highway passing Zone Industrielle on the right, continue westward past impressive clusters of concrete towers of low-cost public housing, to *Pont de la Gabarre,* the bridge that spans the narrow seawater channel Rivière Salée. Rivière Salée separates Grande-Terre—an eastern wing of low-lying limestone, tropical karst hills and of sugar plantations—from Basse-Terre, Guadeloupe's rugged west-ern wing—a mountainous volcanic island dominated by a live, if currently dormant, Mount Soufrière.

At Baie Mahaul turn south on Highway N1 toward Petit Bourg, then 2.4 miles (3.9 kilometers) later turn right to go inland on Highway D23, the highway that crosses Basse-Terre.

Ascend the mountain, passing Vernau on the right, after which the highway bends and winds its way to the intermontane divide. Stop at the *Cascade de Ecrevisses* in this tropical montane forest at a turn-off to the left, near Maison de la Fôret. Continue down the leeward slope of the mountain to Mahaut on the coast. The forest vegetation is luxuriant and thick. Reaching the leeward, Caribbean coast, turn south in Mahaut on Highway N2 and cling to the coastline, winding around headlands into and out of valleys, past tourist beaches to reach Bouillante. In *Bouillante,* stop to visit the geothermal plant, or investigate the beach front and town.

From Bouillante, we continue south along the twisting coast road, winding around headlands, bypassing small fishing villages, such as Petite Anse and Anse a la Barque. Passing through Mari-got, we turn inland and stop at *Beaugendre,* a renovated coffee plantation.

Continuing south, we pass the town of Vieux-Habitants, with its *plage naturiste* (nude beach) and stop at Plage de Rocroy, a black-sand beach. Then travel on to Basse-Terre, the administrative capital of Guadeloupe and its southern commercial port. Passing through Basse-Terre, we linger to comment upon the architectural styles of the government buildings, and the mix of hotels and buildings along the shoreline boulevard.

Leaving Basse-Terre we swing inland past Fort St. Charles, continuing on Highway N2 through suburban Basse-Terre, by-passing Gourbeyre. At Dos d'Ane, we turn right, off the Highway N2, and take the old N2 to *Trois Rivières.* At Trois Rivières, visit the *Parc Archéologique des Roches Gravées,* where there are Arawak or Carib carvings of animals and hunting scenes. This is one of the few sites in the Caribbean where archeological evidence of the pre-Columbian inhabitants is found. This is a walk of about half a mile (0.8 kilometer).

Leaving Trois Rivières, join Highway N1, leading northward up the windward coast of Basse-Terre.

Beyond Petit-Bourg, the circuit is completed when the route turns eastward at Baie Mahault to re-enter Point-à-Pitre.

Point-à-Pitre, one hour and a half

The starting point for this walking tour is Quai 2, immediately opposite the Port d'Autonome de la Guadeloupe offices, where cruise ships tie up. Walk eastward to the old harbor, La Darse, rounding the corner where a reconstructed colonial structure now houses one of Guadeloupe's finest and most select restaurants, La Cannes a Sucre. Emerging at a block of re-created colonial mansions and nouveau-Antillean complexes, we see the Hotel St. Johns and boutique-cum-office complexes on our left. These are immediate physical signs of a comprehensive gentrification and renewal of this once-dilapidated warehouse district of Point-à-Pitre. A few remnants of this past remain, but the dry-goods, merchandising, and storage functions of these wharf-side structures have given way to tourist boutiques, Haitian art galleries, and corporate and insurance offices.

Along Layrie Lardeney, suddenly the smells of the Caribbean invade the senses. Stalls laden with tropical fruits, baskets and straw hats cluster at the head of La Darse. Gone are the rural Guadeloupe hucksters, and these stalls are now operated by Dominicans, the women's familiar straw hats distinguishing them from their local peers, whose colorful Madras cotton turbans are their traditional identification. Cross the Rue Duplessis and enter the Place de la Victoire, a wide park where poinsettias and palms still remain. Here signs of Hurricane Hugo's ferocity are evident in the hewn stumps of what used to be magnificent samaan shade trees.

Turn north and walk through the park, where the mixture of hotels, shops, and buildings on either side provide a profusion of architectural styles rather than classical wholeness. Memorials commemorate the youth of Guadeloupe who died for France in both World Wars. Another memorial commemorates a Gaullist patriot, Felix Eboue, governor of Guyane (French Guiana), who refused to recognize the Vichy government of occupied France. The memorabilia strengthen the sense of Antillean linkages with European France, which pervades these urban societies *d'outre mer*. Across the park on its eastern side stretches the administrative building of the Sous-Prefecteur, this one sporting a new roof after Hurricane Hugo summarily dispensed with its old one.

Turn left and leave the park via Rue A. Isaac, past more three-story colonial buildings. We enter another square, this one cobbled. Here, the triumvirate of metropolitan power and authority is juxtaposed, as if to mimic Latin American plazas and the dictates of the sixteenth- and seventeenth-century Spanish Laws of the Indies. To the south is the Palais de Justice, to the west the National Guard Headquarters, and to the north the Cathedral. Passing on westward along rue Barbes to rue Noizière, we turn south on rue Noizière and, where it joins rue Reymier, come upon the Place du Marché, where stalls selling fruit and a remarkable range of wares catering to *négritude* consumer tastes are crammed together to fill the square. Haitian hucksters, and more Dominicans, appear to have supplanted their rural Guadeloupean counterparts in these street-vending activities in the *Place du Marché*.

Passing the covered market on its north side, we turn north on rue Schoelcher, then east to rue Noiziére and make our way along this main commercial street to the highway. Boulevarde Chanzy, a filled-in canal which was once the northern edge of the old city of Point-à-Pitre, is now a four-lane highway expediting east–west-flowing traffic through the city. The towering new concrete government buildings contrast markedly with the three-story (at most) mixture of individual and different-styled colonial buildings of the old city. Here, north of boulevarde Chanzy, are a few examples of Point-à-Pitre's postmodernist structures. Together with the older monolithic concrete government buildings, they constitute a hub of major architectural symbols of Point-à-Pitre's socialist government and its bureaucratic control. Fifteen years ago, the area north of this drainage canal was a "bidonville," a shantytown.

Continue west along the south side of boulevarde Chanzy where Haitian and Dominican street vendors block shop fronts with their makeshift stalls and sunglasses displays. Then turn south along another of Point-à-Pitre's main commercial streets, rue Friebault. Lebanese dry-goods and fabric stores inter-mingle with *petit béké* commerce—radio, jewelry, and even scooter stores. Pharmacies, on the other hand, are the reserve of Guadeloupean black professionals. This truly Caribbean mix of *comprador-elite* power evinced in rue Friebault is duplicated elsewhere in Caribbean cities. Not unique to these French domains, Caribbean class, color, and ethnicity distinctions parallel other parts of the region, whether Spanish, French, Dutch, or English in colonial roots.

Some distance to the south, the sea front and port barriers mark the return of this walking circuit around the old city. Walk toward the sea and pass many examples of the elegant three-story buildings that characterize the combination of ground-floor commerce and of some interspersed residential upper stories, which were the landmarks of the old commercial sector. Many are in very different states of repair and gentrification. The lasting impression is of regeneration and change in this *vieux cité*. The old has already succumbed to the new.

△ *Day Seven*

LA FRANCITÉ DES PETITES ANTILLES: MARTINIQUE—L'UNICITÉ MARTINIQUAISE

Martinique is separated from Guadeloupe by the island of Dominica, an ex-British colony where many of its people maintain strong ties with one or the other French department, even forming significant illegal communities in Point-à-Pitre and Fort-de-France. Martinique is approximately 40 miles (64 kilometers) by 16 miles (25 kilometers) of rugged and mountainous volcanic rock. The southern hills are less precipitous and the flat alluvial Lamentin Plain immediately to the south of Fort-de-France provides some contrast in scenery. The most infamous of its volcanic peaks, *Mount Pelée* in the northwest of the island, is more often than not enshrouded in clouds. The interior mountainous massif of the *Pitons de Carbets* is a national reserve, and like elsewhere in the volcanic Antilles, the windward and leeward sides of the island differ markedly in precipitation. In the extreme south are cactus and thorn savannas; the rest of the southern peninsula is semi-arid. Extensive tropical deciduous and semi-evergreen forests cloak the interior mountainous slopes, and the arable acreage is only approximately sixty-five percent of the total surface area of Martinique.

Martinique is relatively urbanized; more than half of the island population resides in and around the capital and main port, Fort-de-France. Fort-de-France assumed its preeminent commercial and administrative position after the destruction of St. Pierre by the *nuées ardentes* (very hot, dense clouds of gas) of Mount Pelée in

Martinique

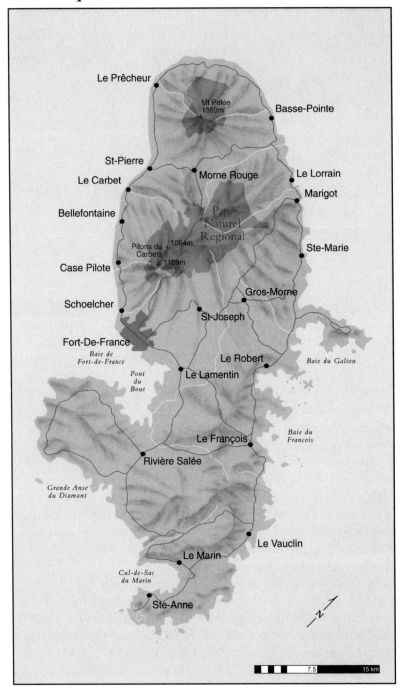

Le Prêcheur

Mt Pelée
1389m

Basse-Pointe

St-Pierre

Le Carbet

Morne Rouge

Le Lorrain

Marigot

Bellefontaine

Parc
Naturel
Regional

Pitons du   + 1064m
Carbets
+ 1189m

Ste-Marie

Case Pilote

Schoelcher

Gros-Morne

St-Joseph

Fort-De-France

*Baie de
Fort-de-France*

Baie du Galion

Le Robert

*Pont
du
Bout*

Le Lamentin

Le François

*Baie du
Francois*

Rivière Salée

*Grande Anse
du Diamant*

Le Vauclin

Le Marin

*Cul-de-Sac
du Marin*

Ste-Anne

N

7.5   15 km

Grande Anse, Martinique, French Antilles. Photograph by Dennis Conway.

1902. Situated on a large enclosed bay deep enough for commercial vessels to dock alongside the quays, Fort-de-France prospered as the district capital of a French department; it prospered as the cultural and educational center of Martinique; the island's prospering tourism and agricultural bases also added to Fort-de-France's growth. Martinique's sugarcane plantations concentrated on producing rum and alcohol distillates rather than bulk sugar for export markets. Her many small distilleries were gradually consolidated as mainland corporations bought out the successful "rhum" brands and capitalized on expanding their market share in Europe. The guaranteed access to French and European markets also prompted the conversion and consolidation of many sugar estates to the production and export of pineapples and bananas. Enclave tourism was another mainland-financed sector, and the *marina* at Pont du Bout across the bay from Fort-de-France reflects a cultural split that is a cautionary sign of tension in Martinique's current society. Here a "France in the tropics" is re-created. The ambiance might be similar to that of Nice or the Corniche; the charcuteries, boulangeries, petits pains, vins blancs, patés et fromages are straight from home (Europe); the hotels, restaurants, marina-bars, auto rentals are modern and international; the clientele are international; only the service and the servers are local, as are, of course, the weather and sea.

St. Pierre and Montagne Pelée

PORT TO ST. PIERRE, FORTY MINUTES

Start at the cruise-ship terminal of the port of Fort-de-France. Leave the port gates and turn left into Avenue Maurice Bishop, a newer road that joins boulevard de General de Gaulle, which is a four-lane divided highway. Almost immediately, turn left to follow Chevalier de Sainte Marthe southward to Fort St. Louis at the southeast corner of La Savane in downtown Fort-de-France. Here we turn westward and our route takes us along the edge of the Baie de Flamandes, where we leave the city via a one-way street, rue Deproge.

Traveling westward on rue Deproge, we cross the Rivière Madame and pass through an industrial and commercial area of the city. We are now driving along Highway N2 (also called rue Attuly). At the traffic lights on the brow of a hill, turn left, continuing on Highway N2 through the suburb of Bellevue. The elevated promontory on the left overlooks the Baie des Flamandes and here clustered together are several educational institutions—Ecole Normale, College Technique, and the Lycée de Jeunnes Filles. Upper middle-class housing can be seen on the right above Highway N2.

Climbing out of Bellevue we reach another wealthy suburb, *Schoelcher,* and five minutes of driving on a smooth wide highway takes us through this upper-class suburban residential area. On the right can be seen townhouse–apartment complexes of community or public origin, a characteristic feature of the comprehensive urban-housing policies of these French West Indies. Schoelcher also houses the Martinique campus of the *University of the French West Indies.* From Schoelcher, the signpost indicates St. Pierre is only 25 kilometers (16 miles) away.

The road winds around the coastline, dropping in and out of small villages—now doubling as commuter-bedroom satellites— such as Fond La Hay and Bourlet. The pattern repeats itself, first a classic hairpin descent into the valley, then a straight road with commercial activities alongside, then another hairpin bend to negotiate as we climb out of the valley. Often the road is cut through volcanic rock, which changes character from ash bluffs to breccia, to volcanic bombs in ash mixtures. The colors vary from dark grays to reddish hues.

Continuing north along this main leeward-coast road we arrive at Case-Pilote. New residential townhouse complexes are evident on the right. Route N2 winds north with mango and flamboyant trees decking the seaward side of the highway and steep cuts on the land sides. Cows tethered to iron stakes graze on the roadside as we pass into another valley where the two chimneys of an oil-fired electricity-generating plant fill the valley scene. This is Fond Laillet EDF, Centrale Electricité de Bellefontaine. Afterward, the road runs along the coastline for a short while, and here is in need of protection from debris falling from the volcanic bluffs high above. A wall and chain-mesh blanket covers the cliff on the right, a narrow beach is on the left.

Now we enter *Bellefontaine,* with its black-boulder beach on the left. Here, brightly painted pirogues (fishing canoes) are lined up together beneath short coconut trees immediately on the left of the road. Bellefontaine contains much older and more traditional buildings than villages nearer to Port-de-France. Many are dilapidated; this village appears to be beyond the capital city's commuter threshold, since there are few signs of townhouse developments and little evidence of urban renewal and gentrification. Leaving Bellefontaine, the road once again moves away from the coastline, into the hills. The hillside vegetation is more xerophytic, though the valleys and some steep gullies or ravines still contain luxuriant forest. The hillsides close to the highway show signs of burning: fresh grass shoots protrude, but volcanic bombs and rocks of lava lie half-exposed above the soil's surface, giving the scene a barren and infertile look. Occasional quarries can be seen, where the volcanic rock has been used as a source of road-repair material.

The newly surfaced highway passes northward through an area of horticulture and varied market gardening—quite a different agricultural scene from the remainder of this leeward landscape. We then enter the village of *Le Carbet,* passing a disused rum distillery, the factory Rhum Meisson. Characteristically, this through-highway runs some way back from the sea front, about three houses deep from the foreshore. Le Carbet turns out to be the only village between Fort-de-France and St. Pierre with speed-bumps to slow traffic for pedestrians.

As we leave Le Carbet, we take the left fork of the highway to take the coast road along the base of steep high cliffs on the right. On the left, brown-to-black sand beaches front the calm and limpid Caribbean Sea. This beach road is reminiscent of the French Corniche, or Riviera. Nets cover the steepest cliffs to prevent rock falls and beaches are found on the immediate left shoulder of the highway. Soon, we pass the *Centre d'Art Musée du Paul Gauguin,* a small but well-designed museum commemorating the sojourn in Martinique of this celebrated impressionist artist, who was better known for his portrayal of life on Tahiti. Then, we pass through a Riviera-like tunnel carved through a headland of volcanic ash, and the bay of St. Pierre opens up ahead.

RHUM J. BALLY

In Le Carbet, we can make a 0.5-mile (0.8-kilometer) detour and visit a working rum factory, Rhum J. Bally at the Plantation de Jus. This entails turning right and going inland toward Morne-Vert, where after about 0.5 mile (0.8 kilometer) the rum factory appears on the left. Many of the buildings are in disrepair, but the factory seems partly to cater to visitors and is partly operational, demonstrating the distilling of rum: Martinique's *"strategie du cannes."* A stop takes approximately fifteen minutes. On our return to the coastal highway, route N2, there are excellent specimens of breadfruit trees along the road. Indeed, the immediate interior of this leeward coast, approximately 0.25 mile (0.4 kilometer) inland from the coast harbors many of these tropical specialties: both here and inland from St. Pierre along the Morne Rouge road.

ST. PIERRE, FORTY MINUTES

Entering the town of *St. Pierre,* the one-time flourishing mercantile capital of Martinique, we veer right to travel northward along rue Victor Hugo. On each side of this one-way street are deep cavernous drainage ditches, testimony to the voluminous surface- and groundwater flows that characterize this city: its precipitation regime and its geologic strata. Rue Victor Hugo is narrow, just sufficient for a bus to pass, but the inhabitants have developed ingenious parking strategies. Having safely negotiated entry of two of their near-side wheels via the few stone covers in place to allow pedestrians access across the ditches, they straddle these ditches.

Stop in the tree-shaded lot of the *Musée Volcanologique* to begin a site tour of the St. Pierre catastrophe of 1902. The museum was created by an American volcanologist, Frank Perret, who later (in

1933) turned over the museum and its impressive collection of photographs, artifacts, and relics to the St. Pierre municipal authorities, for them to manage. The museum is modest in size, but it has an impressive collection of period pictures and glass cases filled with relics and souvenirs of St. Pierre before and after the 1902 eruption of Mount Pelée. The museum does not distribute or sell literature or documentary materials, however; these must be purchased in Fort-de-France. The commentaries accompanying the pictures of St. Pierre at this dramatic and catastrophic turn of the century are in French, but they visually tell a tale of the wholesale destruction wrought by the *nuées ardentes,* as these dense clouds of incandescent gases rolled down the slopes of Mount Pelée and engulfed the whole town and its inhabitants.

From the museum, it is a short walk up rue Victor Hugo to visit two preserved sites, the theater and the prison next door, where the sole recorded survivor, Lucius Sylbaris, was imprisoned, and where the thickness of his cell walls apparently saved him from the *nuées ardentes* on that disastrous day in 1902. The stay in St. Pierre will take about half an hour.

RETURN THROUGH THE INTERIOR, THIRTY-FIVE MINUTES

Leaving the St. Pierre museum, we drive north to the junction where route N2 turns inland and climbs up to Morne Rouge and a minor coastal road continues north to Le Precheur. Our itinerary continues inland to eventually make a circuit to Fort-de-France through Martinique's Parc Naturel Regional along Route N3. This itinerary briefly documents this different circuit and scenic drive through Martinique's majestic interior, a preserved wilderness of luxuriant tropical forest, with a profusion of impressive and sharp-pointed volcanic peaks, topped by the Pitons du Carbet.

Leaving St. Pierre and climbing up the valley, the vegetation exudes tropical verdure, the massive stands of bamboo contrasting with the huge breadfruit trees, with the tree ferns mixed among other broad-leaf tropical species. Driving toward *Morne Rouge,* the domineering presence of a cloud-enshrouded Mont Pelée gradually overshadows the forested scenery on the left. Morne Rouge is a bustling market town, which also shows a few signs of tourist-related activities. Before reaching the city center, however, our return circuit route obviates a right turn onto Route N3. Turning our backs on Mount Pelée and looking toward the mountainous interior of the island, we descend through the suburban outskirts of Morne Rouge. Soon, the forested volcanic cones come closer and on the right there is a spectacular stand of enormous tropical hardwoods, absolutely covered to the crown with epiphytes and elephant ear's vines. Behind this stand of trees are the ubiquitous rows of bananas, many with ripe stems of fruit covered with their characteristic, insect-repelling blue plastic bags. One of the joys of this interior journey, however sinuous the route, is the smooth surface of the highway. Potholes, a standard hazard in the Caribbean, especially in rainy regions, are rare in Martinique. The tropical verdure is almost excessive in its impressiveness.

Driving through the heart of the Parc Naturel, the forested volcanic cones are the backdrop to a profusion of a variety of tropical rain-forest species: Here, on the right-hand side, are banks of chaconia, masses of bamboo, mango trees, and tree ferns. Epiphytes cling to the trees; lianas fill the middle ranges. There are several canopies at differing heights where this route cuts through the montane forest. The winding route sometimes narrows to zigzag and cross small-bridged streams, where a sharp horn toot warns approaching traffic of your "*priorité*." Cars park at various springs along the side of this route, where bamboo spouts protrude from volcanic rock bluffs. Here water bottles are filled and salt-covered cars are occasionally washed. There is river bathing at the Alma crossing. A smoking charcoal pit exudes blue smoke farther along the road, but in general, to this point the forested interior even along this major route is devoid of habitation. Indeed, in contrast to many other Caribbean islands, there is a noticeable lack of habitation and small farming settlements in the interior of rural Martinique. Only when we draw closer to the outskirts of Fort-de-France, as if within the commuting threshold of the city, does settlement begin to appear, and the roadside exhibits its usual Caribbean mix of dwellings, rum shops, car- and tire-repair shops, and the like.

Now descending, we pass *Plon Anthurium* and the Arboretum pull-off, and signs of flowers and plants for sale hang from gateposts. The drive through the Parc Naturel takes only half an hour, a tribute to the smooth road. Eventually, the vista opens to reveal the Baie de Fort-de-France ahead. The highway winds its ways down the tops of ridges with extremely deep ravines to the sides. Here, on the ridges, are the grand old houses of the Fort-de-France elite, taking advantage of these exposed and windy sites on the crests of these south-oriented ridges. Lower down are inter-spersed more modern three- and four-story townhouse develop-ments among these prestigious neighborhoods, but the roadside locations are completely taken up with residences of the old elite. Quite abruptly, we come upon a domed, mosque-like church, *Sacre Coeur de Montmartre,* on the right. The highway winds its way down through more densely populated residential areas. On the lower slopes, the steepest slopes appear to accommodate shanty dwell-ings in various degrees of upgrading and completeness.

Coming upon a roundabout interchange where Route N3 meets the city's circular highway, and given some luck interpreting the signposts and following the directions through this maze of con-struction, we join the eastbound carriageway of Fort-de-France's new circular highway. Soon along this highway, signs to Centre Ville direct us back to the boulevard de General de Gaulle and we arrive at the center of the city.

Fort-de-France, forty minutes

This short walking tour is an introduction to the city center of Martinique's primate city, Fort-de-France, to demonstrate the chang-ing complexion of a French West Indian commercial center, which now more than ever reflects accommodations to tourists. The informal commercial sector and the domestic-metropolitan retail-ing and tourist services have become ever-increasingly intertwined, and are reflected in the city's varying architectures—public spaces given over to market stalls, maritime commercial establishments sharing their waterfront locations with bureaux des changes, rent-a-car offices, tourist (duty-free) shops, art galleries or Centre des Artes Artisanales with their profusion of Haitian art. *Béké* (elite, white families) money is reflected in franchised, and modern, metropolitan (French) department stores selling perfume, haute couture, clothes, and jewelry for "European tastes." Restaurants and hotels abound, catering to the rich and the not-so-rich mass tourist. Tourist facilities invariably are well decorated in their attempt to lure and tempt the cruise-ship or short-term visitors.

This short walking tour starts at the southeast corner of La Savane at the foot of the promontory that still houses Fort St. Louis. Walking westward along the boulevard Alfassa, and passing the statue commemorating Martinique's corsair "discoverer," Belain d'Esnambuc, there is a covered market largely catering to tourists. We encounter a surprising mixture of itinerate vendors and craft sellers—Dominicans, Rastafarians, Haitians, and the occasional out-of-pocket French student, yacht-bum, global wanderer, or traveler— each with goods to tempt the tourist. There are Haitians with their mass-produced artisanal paintings, native carvings, even Philippine carved wares. The Rastafarians sell their own superbly crafted yet simple leather goods, sandals, caps, purses, carved gourds, and charm bracelets, but also imported and mass-produced Ethiopian beadwork. The Dominicans are selling their straw hats and raffia crafts. The itinerants peddle modest inventories of drug paraphernalia or charm bracelets. Unfortunately, turtleshells and stuffed turtles are added exotic commodities.

We continue along rue Deproge, this time on the sidewalk, and the mélange of commercial and tourist-oriented stores is amply demonstrated in the contrasting signs and evidence of recent conversions of operations: from ship chandler to photography shop, from warehouse to rent-a-car office, or small shopping mall. Spanning no more than a U.S. city block in distance from La Savane, this mixture of commercial enterprise and change abruptly gives way to more traditional retailing and warehousing beyond the intersection with rue de la Republique. On the left and toward the bay, the quay side opens out to become the city's major bus and taxi terminals, with their attendant sandwich shops. However, unlike Point-à-Pitre, there is little sign of street-vendor competition here in this large parking area. Automobiles and buses dominate the area.

Our tour turns north on rue de la Republique, where several original and much-dilapidated commercial warehouses remain, albeit with colorful coats of paint—destined to fade under the tropical sun to pastel colors. Then we are firmly reminded of the current fascination with things "American" in these French West Indies and the penetration of U.S. "tastes" globally, by the sight of a Burger King fast-food restaurant. The garish sign almost obliterates the view of an ancient dome-like palace farther along rue de la Republique, La Madrague. At the intersection with rue Victor Hugo, we look down the street and see traditional three-story commercial buildings, most with elaborate and ornate wrought-iron balustrades and balconies, a few with wooden balconies.

Place Volny, a once-open park, is now filled with temporary stalls, where Martiniquans vie for customers, selling imported négritude fashion —wigs and hairpieces, all kinds of dark leather and vinyl ladies' bags, belts, and scarves. This appears to be a "modern" chic niche market for local suitcase vendors.

Now walk east along rue Moreau de Jonnes, past the dilapidated southern side of the Palais de Justice, and suddenly there is a railed park, the Place de Schoelcher, with the statue of Victor Schoelcher at its center. The facade of the Palais de Justice facing this statue of Martinique's "father of slavery emancipation" is commanding and well preserved. The park, too, is well preserved, with walkways crisscrossing it in cardinal directions. It reminds me of Woodbrook Square in the center of Port of Spain in Trinidad. The buildings catering to tourists—such as hotels, restaurants, and tourist shops— seem to be generally better maintained and given newer, more colorful coats of paint than the public buildings. The latter are invariably in various shades of faded gray; many are mildewed and smudged with faded and cracked stone facades. Hotels and restaurants like the Hotel de Palais Creole and Boulangerie, Patisserie Viennoise, on the other hand, are better maintained.

Turning south on rue Victor Schoelcher, and walking two blocks, we reach the yellow and ochre Cathedral Saint Louis. It looks over Place de Monseigneur Romero—this small park being named after the Salvadoran "martyr," murdered by a right-wing death squad in that troubled Central American republic. Here, the juxtaposition of old, new, and postmodern architectural styles, catering to the metropolitan tastes in Martinique, again strike the eye as we walk around this small square. Then continuing eastward on rue Antoine Siger, we soon reach La Savane.

We cross rue de la Liberté and turn north to walk under the tree-shaded walkway lined with traditional, European park benches, where men sit and discuss the latest political wrangling in Paris or in the prefecteur. Here the occasional tourist family sits on the few benches, which during that time of the early afternoon are sun-drenched and, quite sensibly, remain unoccupied by the locals. First we come upon the marble statue of the Empress Josephine, in considerable disrepair. Continuing past this memorial to Napoleon's Martiniquese wife, dominating the northwest corner of La Savane is the Bibliotheque Schoelcher, its ornate ironwork and central dome providing an elaborate contrast to the surrounding business buildings with their subdued colonial architecture.

Crossing to the west and shaded sided of rue de la Liberté, we complete our short walking circuit of the central city by walking south toward the sea front. This last leg takes us past old hotels, a mildewed central Port Office, the newly, and tastefully, painted Musée Départemental Archéologique, several restaurants and tourist-focused stores, where we arrive at the boulevarde Alfassa and the sea front. Daily streams of tourists and commuters cross the traffic on the boulevarde to embark and disembark at Quai d'Esnambuc; the ferry pier which connects downtown Fort-de-France with Pont du Bout, the major tourist center enclave across the bay, and other peripheral tourist and residential areas such as Anse Mitan and Anse-a-l'Ane.

St. Lucia

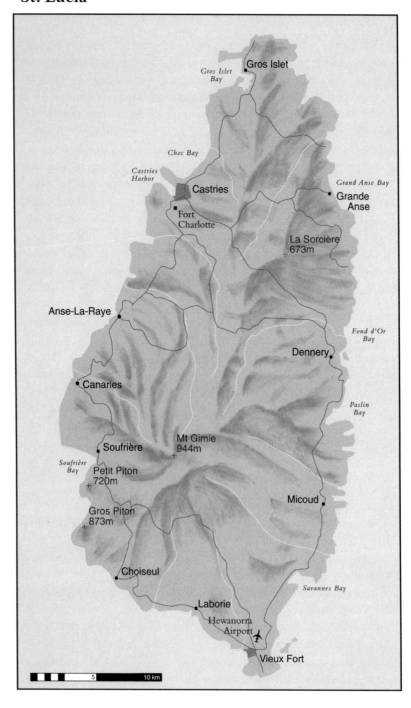

◺ *Day Eight*

ST. LUCIA—APPROPRIATE DEVELOPMENT FOR WHOM?

St. Lucia, fondly referred to by one of her favorite sons, novelist Derek Walcott, as the beautiful "Helen of the West Indies," is one of the largest of the Windward Islands, with an area of 238 square miles (619 square kilometers) and a resident population of approximately 150,000. It is one of the series of mountainous islands of the inner volcanic arc, with the main ridge of its highest mountains forming a north–south backbone for the island. The highest peak is Morne Gimie at 3,117 feet (1,372 meters), but the most scenic and photogenic volcanic peaks are the Pitons—Gros Piton at 2,619 feet (873 meters) and Petit Piton at 2,461 feet (720 meters)—two magnificant pyramids rising majestically out of the Caribbean Sea to the immediate south of the town of Soufrière on the southwest coast of the island. Soufrière is also the location of a far from passive volcano which has a history of rumblings, earth tremors, and emissions, with active solftaras and fumeroles at Sulphur Springs and on nearby hillsides serving as visual reminders of the threat of volcanic activity to the local communities in this part of St. Lucia.

The mountainous terrain and orientation of the backbone contribute to dramatic variations in precipitation in St. Lucia. Annual totals vary from 40 to 150 inches (100 to 375 centimeters) with drier rain-shadow effects reducing totals on the leeward slopes and southeastern coasts, while orographic lifting of the constant easterly Trade Winds contributes to the higher totals in the mountainous interior. The primary tropical forest has been severely depleted by cutting and burning as well as centuries of hardwood logging for construction and boat building, so that most of the interior forest is secondary growth, albeit luxurious where precipitation is high.

St. Lucia was apparently bypassed by Christopher Columbus in 1502, in his fourth voyage to the Caribbean New World. The island was inhabited and successfully defended by the resident Caribs until after the island was settled, and claimed, by the French in 1650, although there are record of previous European landings, such as that by the British in 1605. St. Lucia, like several other Windwards, was a battleground for French and British colonial warring for more than a century, with local Carib tribes often serving as mercenary troops for their European "masters." Eventually, the 1814 Treaty of Paris sealed St. Lucia's fate as a British possession. The island continued to be governed by the British Colonial Office, first as part of the Windwards, then in 1967 as an Associate State, finally gaining political independence in 1979, while remaining within the British Commonwealth. Since political independence, from 1982 to present, a succession of parliamentary governments have been dominated by one premier, Prime Minister John Compton, who has served consecutive terms as leader of the St. Lucian United Workers Party and remains today as the island's senior statesman.

Land under arable cultivation is limited in St. Lucia due to the hilly terrain, as elsewhere in the Windwards, and its distribution is highly unbalanced: A high proportion of small farmers account for a low share in land ownership. Unlike Grenada, where approximately a third of those farms growing bananas on less than an acre are owned and cultivated by small farmers, in St. Lucia more than two thirds of the banana producers are cultivating this "green gold" on such small farms. Originally St. Lucia was a colonial sugar-plantation economy of estates with French–Creole plantocracy, slaves, and, later, estate laborers and small farmers. St. Lucia switched from this excessive reliance on the export of sugar to producing and exporting bananas during the postwar period. In 1961, the fledgling Geest Company bought the Roseau and Cul-de-Sac Valley sugar estates and factories and converted them to banana production, favoring the Lacatan banana over the Gros Michel. The Gros Michel, introduced earlier, was found to be highly susceptible to leaf-spot disease. Today, St. Lucia ranks as the largest producer of bananas among the Windwards, with annual exports of forty percent of the regional total. After more than thirty years of profitable enterprise on its own corporate behalf, Geest Industries still controls the Windwards banana industries. It has long since moved away from estate ownership and today relies on "field-packing" among other cost-cutting measures to minimize its transportation and marketing costs. This saddles the Caribbean small farmers with any overruns on production costs which climatic fluctuations and fluctuations in world market conditions might impose on this international industry.

During the 1980s, St. Lucia has made considerable efforts to diversify its economic base. Annual measures of increases in gross national product per capita gains suggest that the Compton government has been successful in this effort. Tourism has grown tremendously; the industrial estates at Vieux Fort have served as the location for an export-oriented industrial base; bananas continue as the prominent agricultural export; expanding domestic and tourist markets serve as a stimulus for food-crop cultivation. Major investments have been made in developing the island's infrastructure, both in and around Castries, and at Vieux Fort and Gros Islet. Most construction is financed by increased external indebtedness or by a reliance on international loans and grants.

A national debate is under way in St. Lucia, concerning the pace of such "development." Some question the long-term effects of unregulated development initiatives on the country's national heritage, its cultural base, and its people. Change there has been, but what about national sovereignty? What about "Helen's" bounty, her beauty, and her serenity? Nowhere is this debate more heated and contentious than in the argument, some would label it mudslinging, over the Jalousie Plantation project, or the Pitons National Park. The situation, sadly, appears to be an all-too-frequent case of unscrupulous tourism adventurism in the Caribbean, where the government and its interested ministries and its private business sector allies support the Jalousie entrepreneurial venture.

Against this axis is a concerned and impassioned alliance of citizens, the *Star* newspaper, renowned artists, and environmentalists, who argue for the protection of the island's national image, its environmental integrity, and its symbolic heritage: namely the establishment of the Pitons National Park. Remaining neutral is difficult, because at the center of the controversy is Colin Tennant, or Lord Glencannon, recent purchaser of the Jalousie Plantation, who sold out to a shadowy international consortium of financial backers of this enclave tourist facility, the M-Group. This is the same Lord Glencannon who bought the Grenadine island of Mustique from the government of St. Vincent in the 1960s, and then turned it into a fashionable, exclusive residential resort and Caribbean hideaway for some of Britain's rich and famous, among them Princess Margaret and her friends and acquaintances. His plans, or those of the mysterious M-Group, for the Jalousie Resort appear not to be above suspicion, and the exclusionary nature of such a tourist development in the heart of the island's most beautiful landscape certainly is an affront to those who value national sovereignty. "Green tourism," "eco-tourism," and "new tourism" are being promoted as appropriate tourism-styles for many of the small islands of the Windwards and Leewards by local dignitaries of the likes of Eugenia Charles, the Dominican premier, as well as local and overseas academic and environmental communities. St. Lucia's plans to further develop its tourism industry, and to develop a national parks system plan such as Grenada's, might be better served if such enclave-tourism strategies as the Jalousie project are left behind. And recent developments suggest that the St. Lucian government is assuming a more balanced perspective toward encouraging green tourism.

Around the Island, one day

CASTRIES HARBOR TO CANARIES, TWENTY-FIVE MINUTES

Begin at the Point Seraphime Complex where the *Cunard Countess* docks, on the north side of *Castries Harbor.* Drive out of the Seraphime Complex and turn right onto the John Compton High-

Cunard Countess *in St. Lucia Harbor. Photograph by Dennis Conway.*

way to drive into the city's center. You pass the three newly constructed and recently occupied modern government buildings housing the National Development Bank, the National Development Corporation, and the government offices of the Ministry of Finance (the prime minister's portfolio). One of the innovations now evident in 1990 Castries is traffic lights, replacing traffic policemen. Around the harbor and at the crossroads of Jeremie Street, on the left, is an elaborate, ornate, covered market, where outside and inside, rural women sell every kind of fruit and vegetable, along with spices, folk medicines, pots, and pans. A vendor who has occupied the same market spot for over a decade sells handcrafted rattan (cane) furniture at the roadside.

At the light, turn right into Jeremie Street and two blocks later turn left up Bridge Street, past the Post Office through the heart of

Classic colonial mansion, Castries, St. Lucia. Photograph by Dennis Conway.

the shopping center. Continue northward, crossing the Castries River via a very narrow bridge. At the T-junction at the Villa Hotel, you turn left, and go up a steep series of hairpin bends to Morne Road. For the preferred scenic trip over La Toc Road, turn right immediately after crossing the Castries River bridge, and continue past the Geestboat banana loading-dock on the north Quay, Number 3. Here, on shipping days a line of trucks laden with bananas, already boxed, stretches more than half a mile (three quarters of a kilometer) on the right-hand side of the road. On the left we drive past a dense residential settlement of chattel houses—locally referred to as a fishermen's village. Ahead on the promontory is the still-unfinished new hospital. Drive around a couple of hairpin bends, around the headland at the mouth of Castries Harbor, circling left and climbing. The road winds up the Morne and eventually joins the shorter "direct" torturous route. Merge right onto Morne Fortune Road and climb to Fort Charlotte at the top with its preserved colonial fortifications.

Beyond the fort, and coming down the hill toward Cul de Sac Valley and Bay, opening out before us is the Hess Oil transshipment station. This is an early-developed facility that took advantage of the deep-water harbor potential of Cul de Sac Bay (see the 1977 St. Lucia National Plan). Development of the Hess storage and transshipment station was undertaken without any environmental impact analyses, and apparently, extensive leveling of two hills irrevocably destroyed Arawak sites. Continuing down the hill five minutes, past Eudovic Studios, we reach the bottom of Cul de Sac Valley, which is one of St. Lucia's largest banana plantations. Climbing up past the Hess station on the south side of Cul de Sac Valley there is still evidence of the massive earth-moving and extensive landscaping of the terrain in and around the facility, notably a rilled and eroded red-earth quarry face on the left. Up the hill, just past Hess, is the Cul de Sac electricity power station, an oil-burning facility.

Over the brow of this ridge, where small wooden residences fringe the road, and past the Marigot Bay turnoff, we eventually emerge into the Roseau Valley, where another large and expansive banana plantation fills the flat, alluvial, wide-bottomed valley. Drive through the stands of bananas, go up the southern side, and stop at the lay-by, to look back toward the Roseau plantation.

From this lookout point the well-surfaced road winds southward on its leeward journey around headlands and up valleys, eventually reaching *Anse la Raye,* a picturesque fishing village. Beyond Anse la Raye, however, and once we cross the narrow metal bridge at the southern edge of the village on the Canaries Road—Soufrière, a sign reads, is a mere 16 kilometers (10 miles) away—the traveling changes dramatically and bone-shatteringly. Some small porkers, peccaries, might be seen rooting on the banks of the river. All the men we meet walk with their cutlasses as an extension of the right hand.

CANARIES TO SOUFRIÈRE, FORTY-FIVE MINUTES

The road, full of potholes and ridges, rutted and falling away at the edges, makes travel slow and hard. We climb up and away from St. Lucia's leeward coast, then after a wide circle, taking us inland, wind sinuously back toward Canaries, past Anse La Verdure and Anse Jamette. The road cuts into the hillside; the cuts are extremely high and liana-covered. Bananas are grown wherever a flatter patch of soil is available. Yet, signs of habitation along this road are few. The average speed anyone can make on this bumpy road scarcely exceeds 10 to 15 miles per hour (16 to 24 kilometers per hour). Occasionally, on a headland there are a few wooden-frame houses. At *Canaries* (Canaries, pronounced "Canawees" in Kweyole, is named after a particular kind of cooking pot used in these rural parts, not after the Atlantic islands of the same name) Toyota, Mitsubishi and Daihatsu trucks have generally replaced the older and more colorful Bedfords and carry-all lorries of yesteryear.

From Canaries, the road again climbs inland. There are a few more houses and glimpses of farmers and their families terracing the hillside and growing "ground provisions" (yams, sweet potatoes, taros), while mango trees and breadfruit trees, among other luxuriant tropical trees, begin to dominate the hill slopes, replacing the smaller species of the drier leeward slopes.

SOUFRIÈRE TO MOULE-À-CHIQUE, ONE HOUR AND A HALF

Avoiding persistent guides who flag tourists down and offer their services to the volcano, we pass Mount Tabac on the left and begin to see vistas of the Pitons ahead as we descend around the hairpin bends of this still-rutted and pothole-filled road. Eventually we reach the *Soufrière* overlook with the town of Soufrière spreading out across the bay, and Petit Piton, the sharpest of the two volcanic peaks, rising majestically above the town.

In *Soufrière,* continue south through the town, turn left for a couple of blocks, then right to continue south up the hill and out of the town, swinging right up a couple of hairpin bends toward Petit Piton—sign-posted to Vieux Fort. Along the Vieux Fort Road and nearing the sulfur springs and volcano turnoff are young cocoa

Soufrière and the Pitons, St. Lucia. Photograph by Dennis Conway.

plantations and signs of recent diversification efforts to encourage cocoa production in these parts. The entrance to the "drive-in volcano," Mount Soufrière, is on the left. The sulfurous odors, the emissions of steam and sulfurous fumes, the heat generated by the emissions, combine to provide a memorable imprint of live vulcanism in the Windwards.

We emerge out of the sulphur springs back onto the Vieux Fort Road, turning south to continue our tour. We pass more cocoa trees growing where space permits and we arrive at the *Dashene Hotel* with its overlook of the Jalousie Plantation. This "enclave-tourist" development project, destined to accommodate an international clientele of the "rich and famous," commands the forested hollow between the Pitons (an area also designated to be the Pitons National Park) with Petit Piton on the left and Gros Piton on the right. Justifiably, this "development" of one of St. Lucia's national treasures has caused controversy and continues to demonstrate the widely differing views held by some government spokesmen and hoteliers as opposed to environmentalists, artists, and advocates for the preservation and conservation of St. Lucia's national heritage. Leaving Dashene and dropping down to a much better-surfaced road, suddenly the vegetation opens to more settlement, fewer trees, and signs of grazing everywhere.

Here are more open vistas with coconut trees and yam terraces intercropped with tomatoes and greens. Toward Choiseul, goats and cows are staked out along the roadside and in open pasture. Most of the houses here—spread out and scattered—are made of cement block. Turn right on the Reunion loop, pass a chicken farm on the right, and enter *Choiseul,* a little fishing village. The main street is charming with its two-story architecture. Most of the houses are wooden and many have framed upper balconies, but the community-built fish market and cleaning facility is modern.

Leaving Choiseul, drive toward Labourie; the housing is more affluent and sometimes quite modern. Settlement density increases and there are signs of community-development projects, both buildings and agricultural diversification. Pandanus or screw-pine grows on the right-hand side of the road; most are young plants. Continue on the modern and well-surfaced road, bypassing Laborie and through

more and more subdued relief, where tethered goats, black-belly
Barbados sheep, and the occasional cow graze the barely grassed
fields. Here, evidence of these small-time pastoralists' dependence
on the regularity of seasonal rainfall is paramount.

Arriving at Vieux Fort, and driving around the *Hewanorra International Airport,* we pass through the northern outskirts of this
small commercial town and designated regional center. Then, on
the eastern edge of the town a minuscule roundabout heralds the
junction of the west (leeward) and east (windward) roads. Drive
south out of this roundabout, with the Atlantic shoreline to the
immediate left and the reconstruction of Vieux Fort's new deep-
water port facility on the Caribbean side. Head for Moule-à-
Chique, a conservation project designated for preservation, where,
having traversed a dirt road in considerable disrepair, we eventually arrive at the top of Moule-à-Chique, the southernmost tip of
St. Lucia.

From *Moule-à-Chique,* the *Marie Islands,* a bird sanctuary, are
in the immediate foreground. Beyond and northward up the windblown Atlantic coast lies Savannes Bay, with its red mangrove
swamps, a National Trust fish-breeding site. Leave Moule-à-
Chique and pass by Hewannora driving northward on the windward route toward Micoud. Coconut palms fringe the windswept
beaches on the right; inland salt-sprayed and stunted vegetation is
scattered among dilapidated residential settlements on the left. At
Savannes Bay, fish pots (traps) are being built, some in the traditional manner from woven raffia and platted pandanas fibers,
others from chicken wire. Volcanic boulders protrude through the
thin soil on this coastal plain; housing is generally in extremely
poor condition.

MICOUD TO ANSE LA RAYE VALLEY AND RETURN, ONE HOUR

Passing *Micoud,* on the left is a large quarry where the volcanic
ash has been quarried for road building. This east-coast road
linking Hewannora with Castries is being improved and widened
to enable it to carry large trucks. Driving north through Patience
and Mon Repos, banana plantations again appear in the valleys
and on the lower slopes. Here, in contrast to the leeward valleys,

bananas are grown among stands of mature coconut trees. This intercropping of coconuts and bananas appears a common mixture for these windward valleys.

At the *Praslin* canoe-building cooperative, canoes are being built using traditional methods that originated in Arawak times. The builders hollow a log, and fill it with water and boulders to spread it. They then build up the bulwarks and prow with planks, using traditional tools and local woods. When you leave the cooperative, continue north to stop at the Frigate Island National Trust lookout. The National Trust controls access to *Frigate Island,* a breeding haven for the frigate bird. All along this exposed coastal road, the trees are stunted and misshapen by the salt spray and constant wind. We suddenly come upon the *Font D'Or* overlook at the apex of a sharp left-hand turn, and pull over on the right. Before us stretches a beautiful seascape of Atlantic rollers, rocky promontories, with the town of Dennery in the middle distance and the hills of La Sorcière behind. La Sorcière is a Carib legendary vista, where the spirit's face when cloud-covered frowns upon actions or when clear exonerates them.

Our route takes us past the suburbs of Dennery, another small regional commercial center, where we turn inland through a more densely populated area with more verdant vegetation. The more affluent houses are often built entirely on stilt-like concrete pillars when on flat ground. On hillsides the pillars serve as the front supports. Height maximizes access to the breeze, while the stilt-like style extends living space on the lower floor among the pillars. (Trinidadian "Indian" housing is similarly designed.) A series of small nucleated settlements lies along this interior main artery, which crosses the island from windward to leeward coasts. The road winds up to the Barre de L'Isle Ridge, ascending into tropical fern forest along a newly constructed and widened highway. Immediately beyond the crest and on the leeward side of the Barre de L'Isle Ridge, the forest vegetation is at its most luxurious, probably a consequence of orographic lifting effects on the air masses that continue to be lifted beyond the mountaintops. Stands of Honduras pine grow in this protected part of St. Lucia's tropical montane forest.

Eventually, this main highway winds down to the *Anse La Raye Valley,* where rows and rows of bananas again dominate the flat valley bottom on the left and right of the road. Some mango and avocado trees provide windbreaks at the inland head of the Cul de Sac plantation, but, closer to the sea, the bananas stand firm and thick, with the familiar blue bags wrapped around the ripening stems. We complete our circuit around the island and join the western leeward road, turn right, and head north on the road that climbs over the Morne to Castries. If time permits, a visit to Fort Fortune might be a suitable final stop on this island tour.

Grenada

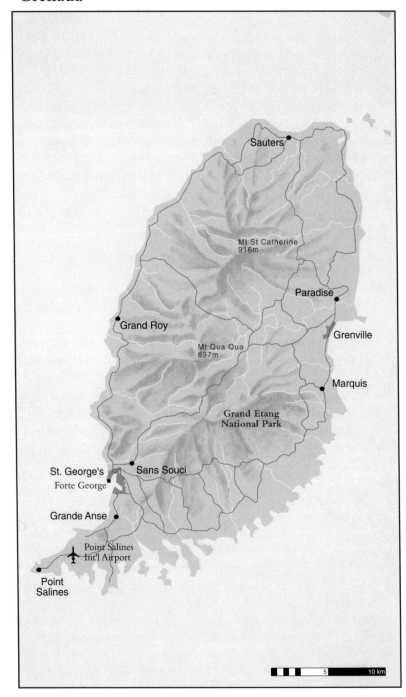

Sauters

Mt St Catherine
916m

Paradise

Grand Roy

Grenville

Mt Qua Qua
697m

Marquis

Grand Etang
National Park

St. George's
Forte George

Sans Souci

Grande Anse

Point Salines
Int'l Airport

Point
Salines

5   10 km

△ *Day Nine*

GRENADA—TIMELESS ISLE OF SPICE

Grenada is renowned as the spice island of the Caribbean. Its fame stems from the profusion of spices it traditionally grew and exported: nutmeg, mace, ginger, cinnamon, and cloves. In addition, the extremely fertile volcanic soils foster the belief that in Grenada "you drop any seed and it grows." Cacao (cocoa), bananas, plantains, paw-paw (papaya), mangoes, and citrus are all favored tree crops on this fertile tropical island. Beneath the tree cover, ground provisions (yams, sweet potatoes, dashine), as well as introduced garden vegetables (tomatoes, beans, melangine), constitute valuable intercropped produce, grown on the small garden plots that inevitably surround Grenadian rural dwellings.

Grenada is the southernmost island of the Windward Island group lying only 90 miles (145 kilometers) north of Trinidad. It is part of the inner arc of volcanic islands, yet one of the smallest in area, a mere 186 square miles (300 square kilometers), a 20-mile by 12-mile (32-kilometer by 19-kilometer) oval. The highest peak in the mountainous interior is Mount St. Catherine in the north, at an elevation of 2,754 feet (916 meters) above sea level. Today, much of this verdant forested interior is protected as a series of national parks set aside for preservation and conservation. The largest, the *Grand Étang National Park,* surrounds a volcanic crater lake and the tropical montane forest of surrounding slopes of *Mount Qua Qua* is jointly managed as a national forest reserve and a national park in line with the 1986 recommendations, which formulated Grenada's forward-looking policies on land management and conservation.

Similar to others in the Windward arc, Grenada's mountainous interior and windward slopes receive high annual precipitation; only the leeward southwest around Point Salines is semiarid. Approximately fifteen percent of the original primal tropical forest still exists, while the majority of forested interior is secondary, yet dense, semi-deciduous and evergreen.

Grenada was settled by the French in the seventeenth century, but after the Treaty of Versailles (1783), the island belonged within the British Empire. Caribs fought incessant wars with both the French and British garrisons. In 1877 the island was declared a British Crown Colony. French Creole planters, white "red-legs" from Barbados, were the estate-owning class, but after emancipation the majority of African-Caribbean ex-slaves and freedmen remained as small farmers. Thus in this isle of spice, a dominant farming peasantry maintained its livelihood and cemented the island's rural traditions. Cocoa and nutmeg marketing boards facilitated the export of these small-farm produce; more recently bananas were grown to supply the weekly rounds of the Geest boats.

In 1974, following abortive efforts to form an economic union with neighboring Trinidad, not to mention considerable political maneuvering on the part of the eccentric premier, Eric Gairy, Grenada was granted political independence from Britain, though it remained a Commonwealth member. Gairy's style of leadership and his increasingly dictatorial behavior prompted popular resentment from these strong-willed islanders. Gairy's strong-arm tactics and recruitment of his own version of Papa Doc Duvalier's Haitian macoutes, the Mongoose Gang, only heightened tensions. Then on 13 March 1979, while Eric Gairy was off the island attending a U.S. conference on unidentified flying objects (UFOs), members of the New Jewel Movement, among them London-trained lawyer Maurice Bishop, responded to the Mongoose Gang's and the "Green Beasts'" overt threats on their lives by toppling the Gairy regime in an almost bloodless coup. For the next three and a half years, Grenada was governed by the People's Revolutionary

Government, before a crisis among the leadership provided an opportunity for intervention by U.S. armed forces in October 1983. Since that troublesome time, rebuilding the Grenadian economy has not been without its setbacks. A hoped-for U.S.-sponsored "Marshall Plan of reconstruction" never materialized. The Point Salines airport was finally completed and is in service, but the largely rhetorical promises for a dynamic rebirth of economic activity after the island was "freed" remain unfulfilled.

St. Georges, the tiny and picturesque capital, houses only about thirty percent of the island's population; the majority still lives dispersed throughout the island, mainly relying on their small farms for the daily necessities. The island administers several smaller island dependencies to the north, Carriacou being the largest. Chartered yachts and cruise ships visit these Grenadine islands, but life here retains its timeless qualities as it does on the mainland.

Grenada has recently changed its celebration of Carnival to the second week in August. This celebrated "Trinidadian" spectacle of calypso (kaiso) and steel-band music, dance, and costumed bands—not to mention unbridled debauchery—usually signals the onset of Lent in February. Long associations between Grenada and neighboring Trinidad have fostered a strong Carnival tradition among Grenadians. Indeed, one of the most, if not the most, famous calypsonian of Trinidad's modern times, "the Mighty Sparrow," is a Grenadian favorite son. Although small scale in contrast to the Port of Spain season of Carnival activities and entertainment, Grenada's Carnival activities constitute a relatively new, but nevertheless informative cultural mirror.

Grand Étang and Mount Qua Qua National Park, a half day

This visit to the forested interior of Grenada consists of a short ride from the cruise-ship berth in St. Georges Harbor to Grand Étang National Park headquarters at the crater lake, followed by a two-

The Carenage, St. Georges, Grenada. Photograph by Dennis Conway.

and-a-half-hour walk through the mountain forest to the summit of Mount Qua Qua and back. This hike requires strong footwear, rainwear, and light lunch packs.

Leaving the cruise-ship berth on the eastern side of the Carenage harbor, we circle the Carenage, and turn right on Matthews Street, passing several completely burned-out government buildings (burned on April 27, 1990). We turn left to pass through Sendall Tunnel and emerge on the Esplanade, where minibus ranks clutter the sea front, and the old police station is being converted to a Melville House insurance building. From the Esplanade, we proceed past fish vendors, north along this sea wharf, leaving St. Georges via Melville Street. Melville Street continues to the St. Johns River valley, where we turn left to go inland on River Road, passing Queen's Park Racecourse and cricket field on the left-hand side of the river, as we drive into the country. The road at first follows the river on the right, but eventually crosses it via an old, wooden, single-file bridge. We go left at a small roundabout and join the main St. Georges–Grenville Road crossing the mountainous interior of the island and traversing the Grand Étang National Park.

The road begins to climb through Holy Oak where peripheral villages like Sans Souci and Mount Gay extend ribbon-like along either side of the road. Indeed, the roadside is one continuous urbanized ribbon of residential development, extremely varied in age, upkeep, and modernity, including gaily painted concrete-block construction. As we climb the air gets cooler; the mountainsides are forested; the road surface is widened and improved. There are cocoa, breadfruit, banana, mango, and citrus trees and the characteristic mixtures of these fruit trees in people's "yards." Forested mountain views to the left reveal the villages of New Hampshire and Willis in the middle distance. The road passes more villages, like Beaulieu and Vendôme, but there are no identification signs (unlike in Barbados, where every turn has its own signpost and directions). Above Vendôme, habitation recedes and only the occasional wooden house still clings to the side of the mountain road. Everyone has a piece of land, where they grow fruit—mango, avocado, guava, grapefruit—spices (the occasional nutmeg tree with its completely shaded area), and their "ground provisions."

The forest reserve is above Vendôme. Farther on is a forestry nursery, where young saplings are raised. Two ministries control this zone—the Ministry of Forestry and the Ministry of Tourism—each with separate staff, offices, and responsibilities. The gullies are steep on the open side of this mountain road. Often gaudily painted barricades protect exposed roadsides, and the cliff cut through the volcanic rock is covered with ferns. At the crest of the hill, on the border of the parish of St. Andrew, is a lookout to the west coast far below. Within fifteen minutes we arrive at the *Grand Étang National Park Centre* and soon the crater lake can be seen through the trees, some of them Honduran pines. We arrive at the well-preserved site of the Ministries of Forestry and Tourism, with two separate buildings and parking spaces, with the Grand Étang crater lake to the immediate left. Pathways are covered with nutmeg shell. The tourist center is beyond the forestry building. Stumps

FEDON

Fedon, a Grenadian folk hero, led the slave revolt of 1795, holding off (with his followers) the French slave masters and the British administrative garrison forces for over two months. Fedon somehow escaped the ensuing massacre of his band, and although the object of a massive search, he was never found.

of the large trees destroyed by Hurricane Janet in 1955 still remain. A radio transmission station occupies the top of one of the hills.

After visiting the National Park Centre, a short walk around the crater lake is our first introduction to this cool altitude modified zone. Hike through the montane forest to Mount Qua Qua. The "Trail Guide Series," produced by the Grenada National Parks, describes this moderately difficult two-and-a-half-hour hike. You will overlook the crater lake and see tree ferns (or *bois jab* as they are called locally), mosses, and orchids. The trail passes through Elfin Forest, miniaturized by constant wind. The trail is well defined. There are only two major junctions, both to the left: the first is a fire trail and the second and more challenging is the trail to Fedon's camp and Concord Falls.

Three hundred yards (274 meters) past the sharp turnoff to the left brings you to large rock formations that afford a breath-taking view of the northeast of the island, the windward coast, the old airstrip of Pearls, and the sea beyond. Retrace your steps along the trail back to the Park Centre.

St. Georges

St. Georges, Grenada's capital city, was once the administrative capital of the Windward Islands, when all were under direct British colonial rule. This colorful city embraces two extinct volcanic craters, the Lagoon and the Carenage. The former last showed semblances of eruption in 1902 and 1929. Indeed, the latest volcanic activity in these parts was the 1990 active rumbling activity of Kick 'em Jenny, an underwater volcano in the passage between Carriacou and Grenada. Less troubled now, the deep waters of the Lagoon and the Carenage are host to yachts, small inter-island freighters and schooners, cruise ships, and local fishing boats. St. Georges perches on the hills and slopes surrounding these two natural circular harbors.

Our walking tour begins at the cruise-ship terminal on the southeastern side of the *Carenage*. Walk along the quay-side around the Carenage. Small fishing boats are tied up on the quay-side; above on the hills surrounding the harbor, residences densely cluster on the lower slopes. Island schooners load on the quay-side opposite. Trading houses, commercial agencies, commission agents, even a betting shop grace the harbor front. Most of the buildings are two stories. Here is the Geest Industries, the British West India Tobacco Company offices, and Grentel, the new telephone company. Some classic British telephone kiosks can be seen on the quay. The Ministries of Tourism, Civil Aviation, and Women's Affairs are next to the Tourist Office. Here, too, is the retail outlet of Spice Island Perfumes, started as a joint venture between the French government and the short-lived People's Revolutionary Government of Grenada. Schooners are loading supplies and passengers traveling to the Grenadines not far from a Daihatsu dealership and traditional family-owned trading companies, such as Ottway, Huggins and Company, and Julian and Company.

Walking on the northern side of the Carenage, we can see alleys on the right where pedestrians can walk up the hill, but the majority of the roads in the town are one-way for vehicles. Tied up on the quay outside the Nutmeg restaurant are rubber dinghies from the yacht basin in the Lagoon. At the corner of Market Street and Young Street is the Department of Inland Revenue, next to several gutted buildings. Across the bay to the left are water taxis. Continue past to Matthew Street and right at M Street to walk up to Sendall Tunnel. The tunnel links the Carenage with the *Esplanade* built in 1894. Returning to Young Street we pass some of the old colonial buildings now undergoing gentrification. One of these was the first hotel in the city, the Hotel Antilles. The steep roads have deep gullies on either side. We go up the hill to the traffic policeman, turn left on Church Street and go up Fort George hill. We pass the Scottish Kirk, or St. Andrew's Kirk, on the right, pass the St. James Hotel on the left, and go up the steep incline to Fort George and the city's central hospital and nursing school on the promontory below the fort. The fort was built around 1705 as Fort Royal, after the 1763 Treaty of Paris. After the Treaty of Versailles more forts were built—Forts Matthew, Frederick, and Adolphus on Richmond Hill. Both French and British had to take on the hostile Caribs in the interior.

Tour *Fort George,* now the home of the Grenada National Police Force. Modern buildings are interspersed with dilapidated structures. If you climb the ramparts to the top, you can look down the leeward coast past Grande Anse all the way to Point Salines. To the left are the Lagoon and the Carenage and farther left is the town of St. Georges.

Exit the fort and go down the George Street hill, past the traffic policeman; the steeple of St. George Anglican church is ahead. All of the streets are one-way, and we are walking against the flow of the traffic as we continue along Church Street up the crest past the British High Commission and the Organization of American States. Past the church, to the left is the market center and to the right is the Carenage. On the left a series of modified sedan porches are evident in the renovated houses. We climb the hill to the Houses of Parliament, and the Grenada Supreme Court and various civil courts. Our walk continues up past Presentation College and on to the St. George's cemetery on the promontory where a small bust of Maurice Bishop is among the poorly tended gravestones in the cemetery.

Returning past Presentation College we come to the crossroads with Market Street and turn right to go down the steep hill to Grandy Street and the Market Square. We still see cast-iron balconies. At the corner of the square we turn left on Halifax Street and turn left again on Gore Street, pass the Chefs Castle (the hamburger revolution has reached Grenada), and walk up the steps to rejoin Church Street. The tarmac road gives way to cobbled surface below the steps. At the top of the steps we take a right on Market Street then veer left past Simmonds Alley, a steep set of steps, on the right, pass round the headland, and once more overlook the Carenage. Where the road stops we turn right, down a flight of steps which originally dropped all the way down to the quay-side, but is now cut off by the Huggins warehouse building. Emerging into Scott Street, we turn left and again ascend an incline to a traffic policeman directing the one-way flow through this intersection, from his roost high up in a booth above the roads. The Roman Catholic Cathedral comes into view immediately ahead. On the left we pass the Grenada Cocoa Industry Board Office. This traditional Grenadian export, cocoa, suffered irreparable damage from Hurricane Janet, and afterward gave way to the Geest-served banana industry. At this Market Hill interchange where the policeman is stationed, we turn right and take the lower road, Tyrrel Street, and finally turn right to circle the Carenage, and make our way past the three-story Marryshow House, now the home of the University of the West Indies extramural division of Grenada. Walking this way in this one-way system means having our backs to the oncoming traffic. Tyrrel Street is so narrow that if the traffic is heavy a safer route would be past the new Grentel telephone headquarters back to the Carenage, where we can walk back to the cruise-ship terminal in relative safety.

Trinidad

△ *Day Ten*

TRINIDAD AND TOBAGO— ARE ALL AH WE ONE?

In terms of its physiography, Trinidad is South American rather than Antillean. The same mountain structures that form Trinidad's Northern range of forested mountains continue beyond the Bocas (a series of islands and narrows) to form the coastal range of Venezuela. Tobago, 22 miles (35 kilometers) to the northeast of Trinidad, is a coral cap of the same west–east-folded continental structures. Central and southern Trinidad also have two ranges of west–east-aligned low hills, scarcely exceeding 1,000 feet (305 meters) in elevation, but there are also wide and flat alluvial plains, in part reclaimed from mangrove swamps. The fauna and flora of Trinidad are continental rather than Antillean, and the variety of species reflects the islands' South American connections.

Trinidad was originally densely covered by thick tropical rain forest and still today approximately sixty-five percent of the island remains tree-covered. At a latitude of ten degrees north, this southerly Caribbean island is safe from hurricanes, so it does not experience the destruction of tropical tree species characteristic of island ecosystems to the north. Precipitation is high almost everywhere, except in the extreme western part of the Northern range and the series of small islands linking Trinidad with Venezuela, where sparse xerophytic vegetation flourishes. The daily regime in the summer wet season is characteristically humid tropical, with brilliant sunshine starting the day; then cumulus clouds build through the morning so that heavy showers invariably interrupt the afternoon's activities.

Trinidad was for many years a Spanish settlement and colony. Then under British colonial rule, the island welcomed French planters and their slaves from neighboring revolutionary French islands. Continuing under British administration, the sugar- and cocoa-estate labor shortages following emancipation of the slaves prompted a search for indentured labor. Chinese, Portuguese, Madeirans, and East Indians (from South Asia) were recruited to work in the plantations, and many remained to mix with others, who immigrated later (voluntarily or otherwise). Among the latter were Levantines (Syrians), Windward islanders, Barbadians, and Venezuelans. Trinidad is veritably a cultural mélange of peoples; "all ah we are one" being a nationalistic phrase invoked to express the cultural diversity that characterizes Trinidad. Superficially, there may be stereotypical landscapes such as the rural Indian and Hindi "South" or the urban, modernizing African-Caribbean "East Main Road," or "Diego Martin," or "white St. Anns." Hindu temples, Muslim mosques, Roman Catholic churches, and Presbyterian chapels signify the country's religious diversity, as do contrasting housing styles. However, North American or Western European models of racial ascription are totally inappropriate means for societal categorization of Trinidad's mixtures. Social class, racial ascription(s), even geography, all play roles in distinguishing among Trinidadians, but today's society defies simple categorization along any one of the many dimensions, or mixed elements.

When sugar reigned supreme in the Caribbean, Tobagonian estates prospered while Trinidadian settlement remained modest. The late nineteenth-century calamities of price downturns, bankruptcies, and bank failures that afflicted many small estate owners, however, caused a collapse of Tobago's wealth and power from which the island has never recovered. Trinidadian fortunes did not wax until the early part of the twentieth century, when British and U.S. companies started to successfully exploit oil in southern Trinidad at Point Fortin and Point-à-Pierre. This southern extractive industry countered the growing importance and administrative primacy of the northern capital and port, Port of Spain. Refineries were established "in south" and the southern city of San Fernando and its county of Victoria prospered both as a regional node and as a symbolic subcultural hearth. Oil wealth combined with sugar wealth, an axis (whether real or potential) of oil-field workers and sugarcane workers posed (or threatened) a unified position to extract advantages for unionized labor; "south" differed from "north." On the other hand, Port of Spain and the conurbation of the county of St. George retained its political preeminence through colonial administrations, through the ill-fated West Indian Federation—where the federation government was housed in Port of Spain—through U.S. "occupation," on to political independence in 1962, and to present day.

South Trinidad, a half day

Greater Port of Spain is an urbanized area housing over half of Trinidad's resident population. The capital city and its Eastern and Western Main Road conurbations constitute one modernizing landscape where three decades of uncontrolled urbanization have left many complex mixtures of housing styles, infrastructural variations, and the juxtaposition of international, American, European, and Caribbean patterns of living, moving, and changing. South Trinidad, on the other hand, reflects its own regional roots and here the East Indian, African-Caribbean, and Creole rural landscapes intermingle with modernizing metropolitan influences less dramatic in their domination and more uneven in their penetration.

PORT OF SPAIN TO CHAGUANAS, ONE HOUR

Our excursion begins at the cruise-ship berth on Kings Dock, in the heart of the city of Port of Spain. Wrightson Road is a divided highway, so we leave the port driving westward past the Holiday Inn on the right; then past some excellent examples of "gingerbread" houses in the Victoria Square neighborhoods, and calypso tents, the Port Authority, and the flour mill on the left. Arriving at the Trinidad and Tobago Electric Power generating plant, we turn right onto Colville Street to begin our circumvention of the city.

Traveling north on Colville Street, we are separated by a wall from the large expanse of the *Lapeyrouse Cemetery* on the right, with its ornate family tombs. Cross Tragarete Street and enter Cipriani Boulevard. The grandiose buildings on the right, for example the National Insurance Board, are mismatched on the left with smaller modest residential structures, now serving as business establishments, restaurants, day-care centers, and the like. This is the edge of *Newtown,* once a residential area, and now an important secondary business node in the city. This boulevard is wider, tree-lined in places, and at its head we join a one-way circuit of traffic moving clockwise around Queen's Park Savanna.

Queen's Park Savanna is a focus for Trinidad sports: horse-racing, cricket, soccer, and rugby all share this wide expanse of grassed commons. Traveling around the Savanna on Queens Park West, the spruced-up appearances of the old Queens Park Hotel and several grand buildings reflect recent renovation efforts to preserve the city's architectural heritage. Rounding the corner to join Maraval Road, a sequence of the finest and most ornate buildings line the road on the left. First comes Queens College, built in German Renaissance style; then "Hayes Court," the residence of Trinidad's Anglican bishop, a Caribbean great house built in a combination of French and English styles; then "Mille Fleurs," a magnificent example of gingerbread fretwork adorning its verandas; next door is the Roodal family mansion, "Roomor House," with its cupolas, towers, and pinnacles and white Italian marble and black French tilework in a French Baroque style; then the Romanesque-styled Roman Catholic archbishop's residence with its copper-sheeted roof and Irish marble and granite construction. The remaining two buildings end this row of magnificent structures: "Whitehall," the prime minister's official residence, is Venetian in style and built of Barbados coral stone; and beyond stands "Killarney" or Stollmeyer's "castle," designed and built in 1904 as a copy of Balmoral Castle in Scotland.

We keep right at the roundabout where the Maraval Road continues northward and are now driving eastward along the northern perimeter of the Savanna. On the left we first come on the *Trinidad Zoo,* then pass the Botanical Gardens and soon see the prime

minister's residence. This was also the ill-fated West Indian Federation prime minister's residence, prior to Trinidad and Tobago's
assumption of political independence in 1962. Ahead overlooking
the Savanna is the Trinidad Hilton. Here, at the St. Anns roundabout, we take the second left turn away from the Savanna up the
hill on the Lady Young Highway. This sinuous highway climbs up
the ridge past the entrance to the Hilton, sweeps around a couple of
hairpin bends to eventually ascend to the ridge-top at the head of
the Cascade and Belmont valleys, the former less densely settled
and falling away to the left; the latter more residentially developed
and stretching out below to the right. At the crest and on the right
we come upon a lay-by, and can pull over for a panoramic view of
the city. Immediately in the foreground is the low-income suburb
of Upper Belmont, and beyond where housing grows denser is the
old middle-class suburb of Belmont proper, the city's earliest
middle-class African-Caribbean neighborhood. The Savanna with
its majestic trees bordering the open-grass expanse fills the view to
the right. In the middle distance the rooftops of government buildings. The hospital and businesses in the northern sector of the
commercial district of Port of Spain mingle with the trees. In the
far distance the "twin towers" which house government offices
and financial institutions and the high-rise Holiday Inn compete
with the flour mill and the electric power facility chimney.

Leaving this lookout spot we descend through the Laventille
Hills and continue eastward along the Lady Young Highway, passing Chinapoo and Trou Macaque "small-island" communities on
the right (Vincentians and Grenadians in the main). The Morvant
Hills are to the left, and these too are unevenly settled with low-income chattel and tapia houses, fronting the highway, clinging to
the steep slopes, and lining up along the ridges beyond. Farther
down the incline, the density of housing increases. On the right, we
pass Success Village, an early attempt by the government to provide low-cost public housing. Further on, a cluster of concrete,
high-rise apartment blocks can be seen on the right; these too were
government attempts to rehouse displaced residents from the
East Dry River Redevelopment project in the early 1970s. On
the left is an industrial estate where the Neal and Massey Group's

Nissan/Datsun plant dominates this commercial landscape. On the lower slopes of Laventille, and entering the Eastern Main Road conurbation, there is a haphazard mix of residential and commercial land uses. This is Barataria. Our route takes us straight over the traffic lights, and at the roundabout where one of Trinidad's latest shopping malls dominates the left-hand side, we join the Beetham Highway as it emerges from the city proper and swoops over an overpass to merge with the southbound traffic on this eastern high-speed conduit of cars and trucks.

This is the Churchill–Roosevelt Highway, the island's first major four-lane highway. The mangrove of the *Caroni Swamp* stretches away to the sea on the right-hand side. Immediately on both sides of the highway are intensively cultivated market gardens. Then we come upon one of Trinidad's earlier industrial estates with Canning's Bottling Company still occupying its original site on the left-hand side. On the right we pass *El Socorro,* now with industrial units mixed with middle-income residences. Twenty years ago, El Socorro was a low-income "Indian" settlement; a settlement of stilt-supported houses among coconut palms reclaimed from the Caroni Swamp beyond the Churchill–Roosevelt Highway. All along the highway are itinerate (now quite permanent) stalls where vegetables and fish are sold. At the traffic lights and major junction of the east–west Churchill–Roosevelt Highway with the south-bound Princess Margaret Highway (now renamed the Uriah Butler Highway), street vendors ply the waiting traffic with bundles of bodie (a long green bean), melangine (eggplant), tomatoes, newspapers. We turn south toward San Fernando.

The Stag Brewery is on the left; then we come upon rice-paddy fields on the left in these Caroni River flats. This whole region is criss-crossed with drainage ditches, many choked with duckweed. Stretching to the horizon on the left is the Caroni Swamp, the nesting place of Trinidad's scarlet ibis, now further colonized by the thousands and thousands of the island's white egrets or "tickbirds." Fringing the highway, the thin ribbon of houses, commonly on stilts, is some evidence of recent satellite settlement along this important arterial road. In ten minutes we pass a spread-out settlement, which has a mosque on the right. This is *Charleyville.* The housing varies in caliber, but plenty of affluence is evident, with the occasional satellite dish a totem of elevated status. We drive on to *Chaguanas,* where the Uriah Butler Highway gives way to the Sir Solomon Hochoy Highway. Turning off the highway at Chaguanas, we turn right and then right again, then turn left immediately after passing under the highway—before getting embroiled in the one-way traffic system of downtown Chaguanas—to continue south on the old Southern Main Road through the small villages of Chase Village, St. Mary's, McBean, and Calcutta Settlement toward Couva.

CHAGUANAS TO MARABELLA, TWENTY-FIVE MINUTES

We pass Chaguanas's Mid-Center Mall. Once we leave Chaguanas, we are in sugarcane country. At Chase Village we come to a highway T-junction where we veer right toward Couva, continuing southward on this original southern main route through "south." Here in this Chase Village, Carapachima region, are traditional potteries where the ceremonial earthenware vessels for Hindu and Muslim festivals are fired in wood-burning kilns. The local red clay is used, but the market for these wares is national. All along this road there is a succession of small businesses: live poultry (chickens and ducks) businesses, retread shops, welding, electronic equipment and repair shops—very much a common scene in "Indian" Trinidad. Now in *St. Mary's* and beyond the junction with Waterloo Road, the landscape to the right and left is quite flat. There is a large plantation house on the right, with its characteristic windmill and clutch of flamboyant shade trees. In marked contrast, a modern school is being built on the left of this Southern Main Road. Small Hindu shrines with their prayer flags can be seen in the lawns of larger houses as we pass first through Lower and then through Upper *McBean* and on to Calcutta Settlement on the left, which is a very modest shantytown area. The main commercial enterprise is selling live chickens for plucking and cooking. Here in McBean, sugarcane country, we come upon the Trinidad Sugar and General Workers Union Sports Complex on the right, one further reminder of the Trinidadian (and West Indian) fixation on their sports culture. At Isaac Junction, turn right to follow the Southern Main Road into and through the center of Couva. *Couva* is a characteristic Indian village, scarcely changed in twenty years, unlike Chaguanas's modernization. Driving west through the Couva main street, we pass many dilapidated old colonial buildings and several housing government offices. The market is small, but still operating. A modern Roman Catholic church is on the left, a convent school on the right, reflecting this church's missionary presence. After 1.5 miles (2.4 kilometers) through Couva we turn left at Roop's Auto Supply toward Point Lisas to continue our journey south.

Immediately south of the Couva built-up area we move into the Point Lisas regional growth center, with the Point Lisas Housing Developments of middle-class housing on the right and on the left building companies. At California we come to a roundabout and turn right into Point Lisas Estate and Port for a ten-minute tour of this major government-sponsored and -financed enterprise; this is the Point Lisas heavy-industry site and deep-water port. The information and public relations office is on the right. Pass the National Gas Company of Trinidad and Tobago on the right, National Commercial Bank, PLIPDECO House on the Century Complex, Caribbean Methanol Plant, Trinidad and Tobago Methanol Company Ltd., FERTRIN, Fertilizers of Trinidad and Tobago Ltd., and on to a roundabout where we turn left toward the port; Trinidad and Tobago Electric Power Company is on the right. The FERTRIN plant is to the left of the road as we drive into the port. The major derricks and cranes can be seen on the left in the docks. We return to the main road and at the California roundabout turn right to continue our southern journey toward San Fernando.

Through California we continue toward Claxton Bay and on the right see the FEDCHEM complex (originally a T & T/Amoco joint venture), now bought out by another multinational, and renamed HYDRO-AGRI of Trinidad and Tobago. Here the road begins to wind a little, approaching Claxton Bay and getting closer to the shoreline of the Gulf of Paria. There's a scattered rural squatter settlement on the left. At the sign post to Cedar Hill Road we enter *Claxton Bay,* a small coastal village, and the County of Victoria. Continuing south through Claxton Bay you begin to catch glimpses of San Fernando Hill in the distance. As the relief begins to undulate and the road winds around low hills, a stand of mango trees emerges. On the left, a major industrial works, Trinidad Cement Ltd., is identified by its large chimney billowing columns of dense smoke into the bay. On the left we come upon Trintoc refinery, the old Point-a-Pierre Texaco refinery surrounded by residential development, for example, Plaisance Park. At the brow of the hill we look down across a host of storage tanks to the city of San Fernando in the middle distance.

Quite soon we enter the "Indian" town of *Marabella,* now part of the San Fernando greater urban area. Here at the street market, sounds of Indian music filter through the crowds as the vendors sell all kinds of garden produce, fowl, and groceries. Leaving Marabella we reach the Union Park roundabout and continue south joining the San Fernando bypass.

RETURNING THROUGH THE INTERIOR, ONE HOUR AND A HALF

At the traffic lights where the sign welcomes us to San Fernando we veer left to turn northward back along the Solomon Hochoy Highway through central Trinidad. To the right is the modern San Fernando Technical Institute; to the left, the usual perched houses of the outskirts of Marabella.

Instead of taking the Naparima-Mayaro Road, turn left onto the Solomon Hochoy Highway. This four-lane highway passes north and east of the Trintoc refinery. To the right are the Trinidadian houses of the workers, to the left the landscaped "colonial" enclave estate of the refinery. We continue north on the highway to the Preysal/Couva turnoff. At the Couva off-ramp, leave the highway and turn right across the overpass to Preysal. The small village of *Preysal* is surrounded by cane fields and beyond it the narrow road to Gran Couva winding its way inland through more cane, with occasional narrow bridges. Bamboo stands fill drainage ditches along the side of this road and ahead the low-lying hills are spotted with stands of trees and scatterings of houses. Moving into the hills, there are older stands of avocado, mango, and breadfruit.

In Gran Couva, housing is modest. We pass through a series of rural villages: Pepper Village, Brasso, Flanagin Town. There are some concrete-block houses, but many are wooden. Nearing Pepper Village, teak trees fringe the road on the right. We occasionally glimpse an old cocoa plantation house with its gingerbread fretwork decorating the balconies. Here the relief is steeper, the trees more varied, and the vegetation more verdant and rich with epiphytes. We pass mature cocoa plantations and mature orange groves in this very rural part of the Montserrat Hills. The road narrows at bridges and there are signs of land slips. The road winds around hairpin bends and this secondary tropical forest is rich and varied.

At the Brasso junction we turn left toward Flanagin Town; in town we turn left again to descend through the Caparo Valley via the Caparo Valley Brasso Road. Here the scenery opens out; a cattle feed-lot called Lake Side Farm is on the left. We pass Flanagin Town Cemetery on the left; the town is very spread out. This is rural African-Trinidadian country rather than Indian Trinidadian. Crop diversification is evident. Corn is grown, and ground provisions and vegetables.

Caparo evinces considerable modernization—something that was not evident farther inland. As we drop into the lower parts of the valley, there are more stilt-houses. On either side of this lower road between Todd's Road and Longdenville are extensive young orange plantations developed by the Agricultural Development Bank. Soon the landscape flattens out, trees become sparse and plenty of Indian prayer flags evident in the yards of stilt-supported housing around Longdenville. This built-up road from Longdenville onward eventually merges with the built-up urbanized and modern area of Chaguanas. At the Amity Gift and Variety Center, the National Gas Station, and just past Montrose Shopping Center on the left, we turn right up the old southern Main Road toward Cunupia, passing through "Muslim Indian" Trinidad territory. The outskirts of Chaguanas are built up and commercial.

Quite soon and as we near the Caroni River, we come upon upper middle-class and commuter subdivisions here in Caroni County. Market garden produce grows the fields. In Cunupia at the Chin Chin Road, which is an alternative route to Piarco International Airport, we go straight on to pass into sugarcane fields and the Caroni plain. Rice is on the left, sugarcane on the right. When we reach the Caroni–Curepe Road, we turn left to pass south of the town of Caroni. On the right is the Caroni Rum Distillery, on the left several churches and mosques. Signs of industrial enterprises on this southern edge of Caroni are plentiful. Cross the single-line Bailey Bridge and take the right fork across the River Caroni, to eventually reach Valsayn Park, where we merge with the high-speed Churchill-Roosevelt Highway and go westward to the city of Port of Spain. This highway passes through Valsayn Park with its various phases of upper-income residential subdivision development, and on to the Uriah Butler Highway traffic lights where our southern circuit started.

PART THREE

Resources

△ Appendix

AREA, PER CAPITA GNP, POPULATION, AND DENSITIES FOR THE CARIBBEAN: 1991

Country	Area, square miles (square kilometers)	Per capita GNP ($US)	Population (millions)	Person/land density	Phys. density*
Antigua & Barbuda	170 (442)	3,880	.1	377	3,267
Bahamas	5,380 (13,988)	11,370	.3	47	3,700
Barbados	170 (442)	6,370	.3	1,548	2,347
Cuba	44,220 (114,972)	n.a.	10.7	243	831
Dominica	290 (754)	1,670	.1	297	1,500
Dominican Republic	18,810 (48,906)	790	7.3	389	1,150
Grenada	130 (338)	1,900	.1	634	1,861
Guadeloupe	690 (1,794)	n.a.	.3	502	1,900
Haiti	10,710 (27,846)	400	6.3	587	1,755

Country	Area, square miles (square kilometers)	Per capita GNP ($US)	Population (millions)	Person/land density	Phys. density*
Jamaica	4,240 (11,024)	1,260	2.5	587	2,454
Martinique	420 (1,092)	n.a.	.3	812	24,153
Netherlands Antilles	300 (780)	n.a.	.2	610	8,275
Puerto Rico	3,440 (8,944)	6,010	3.3	959	6,400
St. Kitts-Nevis	140 (364)	2,860	.04	288	923
St. Lucia	240 (624)	1,810	.2	639	1,548
St. Vincent & the Grenadines	150 (390)	1,200	.1	760	1,334
Trinidad & Tobago	1,980 (5,148)	3,160	1.3	649	2,116
France	211,210 (549,149)	17,830	56.7	268	774
The Netherlands	14,410 (37,466)	16,010	15.0	1,044	4,409
Spain	194,900 (506,740)	9,150	39.0	200	488
United Kingdom	94,530 (245,778)	14,750	57.5	609	2,072
United States	3,615,100 (9,399,260)	21,100	252.8	70	337

Sources: Population Reference Bureau, *1987 and 1991 World Population Data Sheets*, Washington, D.C., 1987 and 1991.

n.a. = not available

*Physiological Density = number of people per square mile of agricultural land for 1987.

△ Hints to the Traveler

EXPENSES AND WAYS TO ECONOMIZE

The Caribbean is not an inexpensive place to visit. Except for a limited amount of domestic agricultural produce, food normally costs 50 to 100 percent more there than in the United States. Even housing is more expensive in the islands than in most of the United States, except for some large U.S. cities such as Los Angeles, San Francisco, New York, and Washington, D.C. On many of the islands car rentals cost twice what they do in the United States and Europe.

The tourist to the West Indies should not regard these expenses as being representative of price gouging because there are reasons why costs are high in this part of the world. Except for a few hardy souls who sail by private yacht, virtually everyone who travels to the Caribbean arrives either by air or cruise ship, the two most expensive ways to travel. Thus, the Caribbean is not cheap to get to. Added to this is the fact that much of what can be purchased in the islands (even much of the food) is imported from elsewhere, so its cost too includes additional high transportation expenses. Finally, much of the tourist industry in the Caribbean is controlled by foreign interests who want a cut of the profits, in addition to the money that is earned by local residents.

The good news is that there are several ways of minimizing the costs of visiting the Caribbean for the budget-minded traveler. First, hotel costs and even air travel are affected by seasonal fluctuations. The highest-cost season usually occurs between mid-December and mid-April and corresponds with the most severe periods of the winter season in North America and Europe. The prospective traveler can save between 20 and 60 percent on hotel rooms by avoiding these dates, especially if willing to vacation during the summer months of June, July, and August. The cruise ship costs and airfares are also substantially less during this period. As noted earlier in the section of this book dealing with the climate in the Caribbean, the summers are not uncomfortably hot as is often supposed. The ocean winds provide a cooling effect for coastal locations where most resort facilities are located. Darwin Porter, a noted travel author, suggests that there are number of advantages to traveling to the Caribbean during the off-season months from mid-April until mid-December, in addition to cost savings. During this period the beaches, hotels, and everything else is less crowded on the islands. Generally, life is less hurried then. Accommodations are easier to obtain. Resort boutiques usually have summer sales to reduce their inventory that did not sell during the previous high season. Reservations for restaurants are less often necessary and their service is usually better with fewer customers. Waiting for a rental car, tennis courts, and to play golf is not as common. In addition, the atmosphere is more cosmopolitan during the off-season because of the influx of cost-conscious Europeans. Even most West Indians prefer to travel among the islands during the off-season.

In addition to season of travel and the obvious effect of length of stay, island destination is a factor affecting both air and cruise ship costs. Clearly, there is a distance factor in these costs, with travel from the United States to the Bahamas being less expensive than a trip to Puerto Rico, and the latter being less expensive than one to Barbados or Trinidad. Furthermore, some of the smaller Caribbean islands do not have direct connections to gateway cities in the United States or Europe. For these islands, it is necessary to fly to the nearest island with an international airport and then catch a flight on a smaller plane to the final destination, adding to the vacation's cost.

Rooming accommodations clearly also affect the price of a Caribbean vacation. The most expensive are found in the large resorts and luxury hotels. There are, however, many modestly-priced accommodations available in smaller, less pretentious hotels and guest houses. The latter are a tradition in the West Indies. Guest houses provide rooms at very reasonable prices and offer an alternative to more expensive accommodations. Sometimes they are located in private homes and some are more like a small simple motel which may be built around a pool. They range in price from modest to cheap, with some being very austerely furnished and the renter having to share bathroom facilities with other tenants. But others are surprisingly comfortable, with air conditioning and private baths. Staying in a guest house often provides a better view of how West Indians live, although some are owned by foreign families who have moved to the islands. When West Indians travel they usually prefer guest house accommodations.

CLIMATE AND APPROPRIATE CLOTHING

As discussed earlier in the climate and weather section of this book, warm tropical conditions prevail throughout the Caribbean. Although variety does exist, the rainy season normally lasts from May through October and the official hurricane season is from June 1st to November 30th. Except for hurricanes, rain has very little effect on tourist activity because the precipitation is normally of short duration. It is advisable to carry a small umbrella, but even this is not absolutely necessary. There is no need to be encumbered with a bulky raincoat. Two items you certainly should carry with you as an adjustment to the intensity of the sun are a hat and sunscreen.

Because of the warm climate, dress standards in the Caribbean are normally casual. Although some businessmen wear coats and ties, most wear slacks, sport shirts, and *guayaberas* (long- or short-sleeve shirts, usually made of cotton with intricate lace work on the front). Some of the more elegant clubs, restaurants, and casinos require sport coats and ties for men and cocktail dresses for women, but most do not.

If you are traveling on a cruise ship it is important to remember that most have a dress code for at least some of their meals. Some require formal attire, or at least a coat and tie for some dinners. But these requirements vary among the cruise lines, so it is best to check with the line you are traveling with ahead of time.

When traveling on an island it is appropriate for tourists to wear almost any type of vacation or outdoor clothing. Short-sleeve shirts, a wide-brimmed hat, shorts or comfortable slacks, and comfortable walking shoes or tennis shoes are recommended. Swimming suits, however, are not appropriate for walking around town or the countryside, and should only be worn on the beach. Many islanders are deeply religious and are offended by tourists who wear bikinis on shopping trips in town. Bathing nude or topless is fashionable at some beaches in the French West Indies, but is still unwelcome on most beaches on the other islands, except where certain resorts cater to a more hedonistic clientele.

HEALTH, HAZARDS, SAFETY, AND THE WATER

Common-sense precautions against health problems during a trip to the Caribbean include taking along extra prescription pills for any illnesses, carrying an extra pair of contact lenses or glasses, and carrying a card or tag identifying any medical condition a person may have. Tourists should also be aware and alert to the possibilities of crime in the West Indies. It is advisable to wear a moneybelt and much of your money should be kept in traveler's checks. Always lock your hotel room, even if you are in it. Crime is on the rise in the West Indies, but it is certainly no worse than in many downtown areas of large cities in North America and Europe.

In addition to infrequent hurricanes, there are a few other hazards that the traveler should be warned about. Driving on the islands can be treacherous because of roads that are generally not as well maintained as those in wealthier countries. On a number of islands driving is on the left side, further complicating the situation.

Many beaches and swimming pools do not have lifeguards on duty all the time. In addition, swimmers should be very careful in bathing at unattended beaches because of the undertows and riptides that are common to many island beaches. Before swimming be certain to inquire about such conditions from someone who knows the beach. Do not dive into unknown waters because of the possibility of rock or coral formations being located near the surface. Also, be careful not to step on sea urchins when wading in tidal pools. It is always advisable to wear tennis shoes when wading or walking along rocky coastal areas.

In most hotels tap water is safe to drink, except in Haiti and parts of the Dominican Republic. However, drinking from freshwater streams is not advisable because of possible pollution. Also, swimming in island streams is not advised because of the possibility of being infected by schistosomiasis, a debilitating disease carried by a parasite that is most often found in slow-moving water, especially on St. Lucia and Puerto Rico.

Poisonous snakes are rare and only a minor problem on the islands of St. Lucia, Martinique, and Trinidad. Unfortunately, all three of these islands have the venomous Fer de Lance, whose bite can be fatal. More dangerous to naturalists and hikers, however, are marijuana farmers. They grow their crop in isolated areas, and assume that outsiders have either come to steal it or to report its presence to local authorities. Many carry guns and some even set trap guns near their fields to scare off the unwary interloper. As a consequence, hiking in the mountains is advised only with a guide or along well-marked and well-traveled paths.

Diseases such as yellow fever, smallpox, and malaria are no longer a problem in the West Indies, except for Haiti and adjacent parts of the Dominican Republic. AIDS (acquired immune deficiency syndrome), however, is on the rise. It is found among the same "high risk" groups (gays, drug addicts, and prostitutes) that it is in North America and Europe, but is becoming increasingly common among heterosexual men and women who engage in casual sex. Sexually liberated men and women should certainly include condoms in packing for their Caribbean vacation, as they should for a vacation anywhere else in the world.

DOCUMENTS AND CUSTOMS

Immigration requirements vary from island to island in the West Indies and according to country of origin for arriving tourists. In the case of arrivals from the United States and Canada a passport is usually sufficient. But occasionally a visa is also required. Sometimes, especially for cruise-ship arrivals, a tourist can gain entry by merely showing proof of U.S. or Canadian citizenship (such as displaying a voter's registration card or an official birth certificate). Still, it is highly recommended that all travelers to the Caribbean obtain and carry passports. They also should check with their travel agents to make certain that they can satisfy all entry requirements for all the islands they plan to visit.

All items brought back to the United States and Canada must be declared to customs agents. Residents of the United States are allowed a $400 exemption for items purchased in the Caribbean and brought home, if they have been outside the country for more than forty-eight hours. American tourists returning from the U.S. Virgin Islands are allowed a double exemption of $800 and are not subject to a forty-eight-hour minimum stay. A flat duty fee of ten percent is applied to the first $1,000 worth of merchandise in excess of either the $400 or $800 exemption. Be sure to keep all sales slips as proof of the value of items purchased in the West Indies because duties are charged according to fair value estimates. Gifts valuing up to $50 can be mailed duty free back to the United States, as long as they are not being sent to the traveler.

Alcoholic beverages represent a special category of import items. Returning U.S. residents are allowed to bring back only one liter of liquor duty free, except those returning from the U.S. Virgin Islands. Tourists returning from the U.S. Virgin Islands can bring back up to five liters of alcohol free, as long as at least one liter was produced there and the other four were also purchased there.

LANGUAGE AND CURRENCY

Although at least some English is spoken on virtually all the Caribbean islands, it is most clearly understood and easily used in the former British possessions. It is most difficult to converse in English on the French-speaking islands, such as Martinique, Guadeloupe, and Haiti, and on the formerly Hispanic islands of Cuba and the Dominican Republic. Although Puerto Rico was a Spanish colony, it is usually easy to find English-speaking help in the major cities, although the situation is different in some of the remote rural mountainous areas of the island. Almost everybody speaks some English on the Dutch islands. Clearly, historical legacy is the most important factor in determining official language on the individual islands. Thus, the former Spanish possessions use Spanish as their first language; the islands owned by France use French; the Netherlands Antilles officially speak Dutch; and the former and current British possessions and the U.S. Virgin Islands use English.

But the linguistic situation is further complicated by the mixture of languages into local dialects called *créole* or *patois* languages. For instance, Haitian *créole* is a mixture of African words with French and Spanish words and phrases. Furthermore, a *créole* or *patois* spoken on one island is not mutually intelligible with one spoken on another. In Haiti *créole* is more widely understood and spoken than French (the official language), especially in the rural areas. Generally, however, *créole* and *patois* are more often spoken as the vernacular among the poor people on the other islands, and are used much less by the better educated middle and upper classes.

U.S. dollars are widely accepted throughout the Caribbean, even though many of the islands have their own currencies. Eurocurrencies and Canadian dollars are also widely accepted in the major cities of most of the islands. Occasionally, there are exceptions to these rules. For instance, in Jamaica all purchases are required by law to be made in Jamaican dollars. Currency exchange offices allow the tourist to convert foreign money into Jamaican dollars upon their arrival and then reconvert it back into foreign money when they leave the island. On the other islands, the best place to exchange money is in the banks or official money exchanges, rather than the hotels, restaurants, or other businesses where generally lower exchange rates prevail.

Cash advances (in local currencies) through major credit cards are also readily available in banks in larger cities.

Because U.S. dollars are so widely accepted, it is recommended that other foreign currencies be converted into either U.S. cash or traveler's checks before visiting the Caribbean. If you decide to exchange money in the West Indies, check the official exchange rate at home before leaving so you will know what is reasonable. Be very careful of exchanging money on the black market in the Caribbean. Currency violations carry stiff penalties on many of the islands and counterfeiting is a problem, although the black market rates offered may appear to be more generous than official exchange rates.

ELECTRICITY AND TIME ZONES

Most electrical appliances purchased in North America such as shavers, hair dryers, radios, portable television sets, and travel irons operate on 110- to 120-volt, 60-cycle alternating current. Some of the Caribbean islands use 210- to 230- volt, 50-cycle electricity which will burn out most U.S. appliances. Some others use 110- to 127-volt, 50-cycle current. The 50-cycle current causes the appliances to run more slowly than normal, which also can damage them. European plugs are used on the French islands and St. Martin, which will require a converter for U.S. appliances to work. Other islands use the same electrical system as used in

North America. In other words, the Caribbean traveler should check to be certain that the electrical current is compatible with his or her appliance before using it.

A line drawn from north to south through the Caribbean and along the border between Haiti and the Dominican Republic separates the region into eastern and western halves. In the western half, the islands set their time comparable to that of North America's eastern time zone (ETZ). The main territories included in this zone are the Bahamas, Cuba, Jamaica, the Cayman Islands, and Haiti. The rest of the islands set their clocks according the North America's Atlantic time zone (ATZ). Places in the ETZ have the same time as the eastern coast of the United States and are five hours behind time in London. Islands in the ATZ are one hour ahead of time in the ETZ, and four hours behind London's time.

△ Suggested Readings

Readers interested in additional background information on the geography and socio-cultural history of the Caribbean Islands might find some useful perspectives among the following selections:

Anderson, Thomas D., *Geopolitics of the Caribbean,* New York: Praeger, 1984.

Despite its title, this is a short book describing the general geography of the Caribbean. It should be considered must reading for anyone interested in the Caribbean.

Ashdown, Peter, *Caribbean History in Maps,* Kingston, Jamaica: Longman Caribbean, 1979.

This is an excellent historical atlas of the Caribbean designed for secondary schools but also useful to adults who want to gain a better appreciation and understanding of the West Indies.

Atlas for Caribbean Examinations, Kingston, Jamaica: Longman Caribbean, 1991.

This is a general atlas of the Caribbean designed for secondary students attending schools in the English-speaking Caribbean who are studying for their examinations.

Barry, Beth, Tom Wood, Deborah Preusch, *The Other Side of Paradise: Foreign Control in the Caribbean,* New York: Grove Press, 1984.

Although written by American scholars, the perspective of this book is a critical exposition of the exploitative influence of U.S. and European interests in the West Indies.

Blume, Helmut, *Caribbean Islands,* translated by J. Maczewski and A. Norton, London: Longman, 1976.

This English translation of a German geographer's comprehensive coverage of the physical and environmental bases of the Caribbean contains a wealth of detailed information on the islands' variety of natural and human-modified landscapes.

Boswell, Thomas D., "Population and Political Geography of the Present-Day West Indies," in *Middle America: Its Lands and Peoples*, edited by R. C. West and J. P. Augell, 3d ed., Englewood Cliffs, N.J.: Prentice Hall, 1989.
This up-to-date piece on the Caribbean that serves as a valuable introduction to the region's settlement and colorful political history is one of several chapters in a well-known college-level textbook.

Box, Ben, and Sarah Cameron (eds.), *1991 Caribbean Islands Handbook*, Bath, England: Trade & Travel Publications, 1991.
This is an annual publication that provides up-to-date information regarding the political and business climate of the islands in the Caribbean.

Caribbean Junior Social Studies Atlas, London: Macmillan Publishers, 1990.
This is another general atlas displaying the characteristics of the islands in the Caribbean designed for older elementary school and younger secondary school students living in the West Indies.

Caribbean Secondary School Atlas, Kingston, Jamaica: Longman Caribbean, 1984.
This is probably the best-known atlas of the Caribbean islands produced for secondary students studying in the former British islands for their final examinations.

Conway, Dennis, *Tourism and Caribbean Development* (1983); *Grenada-United States Relations, Part I, 1979–1983: A Prelude to Invasion* (1983); *Part II, October 12–27, 1983: Sixteen Days That Shook the Caribbean* (1983); *Trinidad's Mismatched Expectations: Planning and Development Review* (1984), Hanover, N.H.: University Field Staff Reports. *Caribbean Migrants: Opportunistic and Individualistic Sojourners* (1986), Indianapolis: University Field Staff Reports.
These field reports, written while the author lived in and traveled throughout the Caribbean, present views of the region derived from field experience, drawing on Caribbean perspectives by using local sources.

———, "Caribbean International Mobility Traditions," *Boletin de Estudios Latinoamericanos y del Caribe* 46, no. 1 (1989): 17–47.

————, *Small May Be Beautiful, But Is Caribbean Development Possible?* Indianapolis: University Field Staff Reports, 1991.

These more recent writings by Conway further develop ideas on the problems and prospects facing Caribbean development, using a socio-historical perspective to consider the widespread adoption of international mobility as a viable and flexible adaptive strategy among the islanders.

Deere, Carmen Diana (ed.), *In the Shadows of the Sun: Caribbean Development Alternatives and U.S. Policy,* Boulder: Westview Press, 1990.

This is a collective effort of seven scholars who examine what they call the economic decline of the Caribbean during the 1980s and describe alternative development strategies for the 1990s.

Diagram Group, *The Atlas of Central America and the Caribbean,* New York: Macmillan Publishing Company, 1985.

This is an atlas of all of Middle America, including the islands of the Caribbean as well as Mexico and the countries of Central America.

Glasscock, Jean, *The Making of an Island: Sint Maarten & Saint Martin,* Wellesley, Mass.: The Windsor Press, 1985.

This book is a history of the founding, settlement, and development of the island of St. Martin.

Hoefer, Hans, Chris Caldwell, and Tad Ames, *Puerto Rico,* Singapore: Hofer Press, 1990.

This is part of a series of books dealing with popular tourist destinations. It provides a general description of Puerto Rico covering a wide range of topics such as the island's physical geography, history, and culture.

James, Preston E., and Clarence W. Minkel, *Latin America,* New York: John Wiley, 1986.

This is an updated version of a classic work describing the general geography of all the countries of Latin America and the Caribbean.

London, Norrel, *Principles of Caribbean Geography: A CXC Approach,* Kingston, Jamaica: Longman Caribbean, 1991.

This is a secondary-level text that introduces the geographical characteristics of the Caribbean. It is designed for students studying for their examinations in the British West Indies.

Lowenthal, David, *West Indian Societies,* London: Oxford University Press, 1972.
This is a classic study of the cultural patterns in the British West Indies.

Macpherson, John, *Caribbean Lands,* Kingston, Jamaica: Longman Caribbean, 1990.
This is an updated edition of a geography textbook for secondary students studying in the Caribbean.

Morrissey, Michael, and David Barker, *Introducing Caribbean Geography (Books 1 and 2),* Kingston, Jamaica: Longman Caribbean, 1990.
These are two relatively new textbooks describing and explaining the basic geographical patterns of the West Indies. They are written by two highly-respected professors from the University of the West Indies in Kingston, Jamaica.

Segal, Aaron L. (ed.), *Population Policies in the Caribbean,* Lexington, Mass.: Lexington Books, 1975.
This is a classic work that describes population problems and policies in the Caribbean. Each of its chapters covers an individual Caribbean country. Although somewhat dated, it still serves as a useful reference for a person with beginning interests in the population problems of the West Indies.

Sealey, Neil, *CXC Geography,* London: Cambridge University Press, 1992.
This is a very recent book covering the principals of geography, with reference to examples in the Caribbean. It is designed for secondary school students studying in the British Caribbean.

Sealey, Neil, *Natural Resources in the Caribbean,* London: Hodder and Stoughton, 1986.
This is a short book written for secondary school geography students dealing with the natural resources of the West Indies and their conservation.

Slesin, Suzanne, et al., *Caribbean Style,* New York: Clarkson Potter, 1985.
This well-illustrated guide to vernacular and local building styles and forms is a mine of information on the French, British, and U.S. architectural influences on the housing stock of the region.

Taylor, Jeremy (ed.), *The Caribbean Handbook 1991*, St. John's, Antigua: FT Caribbean, 1991.

This is an outstanding treatment of the main tourist sites in the Caribbean. Over 600 pages long, it is clearly the most comprehensive guide to the West Indies.

Watts, David, *The West Indies: Patterns of Development, Culture and Environmental Change Since 1492*, New York: Cambridge University Press, 1987.

This is a beautiful historical geography treatment of the Caribbean beginning with the Indians living in the West Indies prior to the arrival of Columbus and ending with the present. It emphasizes the relationships between humans and the environment in the West Indies.

West, Robert C., and John P. Augelli (eds.), *Middle America: Its Lands and Peoples*, Englewood Cliffs, N.J.: Prentice Hall, 1989.

This is probably the best geographical treatment of Middle America, including the islands of the Caribbean, the countries of Central America, and the islands of the West Indies.

⚠ Index